The Biodynamic Movement in Britain

A History of the First 100 Years

BERNARD JARMAN

To Katherine Castelliz for her inspiration; my mother, who encouraged my love of nature as a child; and to the many wonderful people in this country and across the world who dedicate themselves to biodynamic work.

First published by Floris Books in 2024

www.florisbooks.co.uk

Also available as an eBook

British Library CIP Data available
ISBN 978-178250-869-4

The Biodynamic Movement in Britain

Contents

Introduction

It is now nearly a hundred years since Rudolf Steiner gave the eight lectures of the Agriculture Course at Koberwitz (now Kobierzyce) in present-day Poland in 1924. At that time, the European world was gradually coming to terms with the aftermath of the Great War and the dawn of a new transatlantic cultural focus. It was also recognising how the umbilical link with an unbroken stream of tradition was being irrevocably severed: industrialisation and the beginnings of today's consumer culture were sweeping away earlier, more sustainable lifestyles. What for centuries had been passed on almost unchanged as a cultural inheritance was now coming to an end. Those who lived off the land felt increasingly estranged from it. They could no longer rely on it in the same way as they had before. The noble and wisdom-filled folk culture that had been the foundation of European civilisation was rapidly and perceptibly slipping away.

The Agriculture Course came about as a result of the persistent efforts of Count Carl Wilhelm von Keyserlingk, the owner of a large farming estate in the former German province of Silesia. Along with several other agriculturalists he had grown increasingly concerned by the decline in seed vitality, livestock health and food quality. This was long before the full consequences of inorganic agriculture, with its heavy reliance on chemical fertilisers and pesticides, was felt. In response to this overall deterioration, Count Keyserlingk approached Rudolf Steiner and petitioned him to provide a new direction for agriculture.

The agricultural lectures were given primarily to those who were involved practically with agriculture. This meant they were viewed as more than just pure theory, but as something to be acted upon. Although their content was, and remains for many, difficult to grasp, the lectures have nevertheless proven to be a great inspiration. Their lasting message is one of hope and faith in the future. Farmers tend to be traditionalists by nature, living as they do with the certainties of the past and the vagaries of the present, and to survive they must hold on to their past. The inspiration provided by the Agriculture Course, however, has also enabled biodynamic farmers to take hold of the future and to keep faith with what will one day become possible. A new folk wisdom is slowly emerging, which is orientated less to what has been and more to what is yet to develop. Biodynamic agriculture in this way belongs to a movement for the future.

Much has happened since 1924. Biodynamic agriculture has developed and spread across the world. More than 618,000 acres (250,000 hectares) of land are now being managed using biodynamic methods on all continents and in the most diverse climatic and cultural conditions. From the simple indications given by Rudolf Steiner, a well-researched science of life has gradually evolved. Techniques have been developed that have, for instance, allowed the biodynamic preparations to be far more consistently and thoroughly applied than had at first been thought possible.

Like the effect of homeopathic and herbal remedies on the human organism, these preparations serve to strengthen the intrinsic vitality of the whole farm or garden organism, and enable the plant to regulate and balance its own internal metabolism, draw in what it needs from its super and subterrestrial surroundings, and enhance its qualitative attributes. These preparations allow the biodynamic practitioner to influence not only the material substances and nutrients required, but also the living processes active in the soil, plants and animals. They also increase sensitivity towards the archetypal formative and harmonising influences of the stars and the rhythmic movements of the sun, moon and planets.

Farming today is suffering a deep existential crisis. Small farms are no longer economically viable and large ones survive through subsidies and an increasing reliance on agrochemicals and biotechnology. That this is the case is due in no small part to the economic system under which we are living. There is no alternative we are told, and yet our current system can only succeed through the fact that it exploits the environment, primary producers (farmers) and, increasingly, low- and middle-income countries. Our economic system, which bases itself on competition and the pursuit of personal profit, has a direct parallel to biology and the theory of evolution. Darwin's theory, so deeply entrenched in our culture, is based on the assumption that existence is ultimately a battle for survival between competing organisms. A new and very different approach to our understanding of the earth and its evolution is rapidly gaining credence, however. If this approach were to be translated into our social and economic affairs it would bring a far-reaching transformation and healing of our troubled society. Instead of the maximisation of personal gain, a new gesture of service and mutuality would come about. This new approach is one in which the whole earth is conceived of as a living, self-regulating organism in which each species and every part of the earth is recognised as existing for the benefit and greater well-being of the whole.

The suggestion made in the Agriculture Course that a farm should be conceived of as a self-contained individuality relates to this and is arguably the most fundamental precept of biodynamic agriculture. Relying as far as possible on home-produced compost and manure for its fertility and home-grown feed for its livestock, each part of such a self-contained and evolving organism (livestock, crops or soil) supports and interrelates with every other for the greater benefit of the whole farm. It then follows that only so much livestock may be kept as can be fed from the farm and only so much land allocated to cash crops as is not required to grow forage. The greatest possible diversity of wild flora, fauna, natural habitats and biozones, as well as farm crops and livestock, is also encouraged. The farm organism then develops its own resilient identity, builds long-term fertility and can provide wholesome produce of high quality.

Another fundamental idea is that as a living entity the farm reflects the organism of our whole planetary system or, in the words of Rudolf Steiner, 'everything which happens on the earth is but a reflection of what is taking place in the cosmos'.[1] This idea also accords with Goethe's discovery that the whole is reflected in the part: 'If you would seek comfort in the whole, you must learn to discover the whole in the smallest part.'[2] This understanding offers an opportunity for working consciously and creatively within the context of the living earth as a whole. A living organism necessarily differentiates its functions within itself and focuses activity by developing different organs. It also exists within a clearly defined yet porous boundary. As part of a greater whole, each organism is then mutually dependent on every other. Much can be learned from this observation of a farm organism that is of benefit for other spheres of life too. The extent and breadth of biodynamic agriculture is such that there is hardly a sphere of life untouched by it.

The journey towards awakening a biodynamic understanding of nature and the way we grow our food, can begin by looking at how our predecessors cared for the land, how farmers, crofters and smallholders related to the natural world in the centuries before our modern age began. Their knowledge, handed down through generations, came from the land itself and the need to make a living from it. Their life with the land gave them an intuitive knowledge of their crops, livestock and soil, and of the weather conditions they had to work with. Their experience of the seasons was permeated by a social culture of festivals and celebrations of religious devotion.

In the western world, little of this old farming wisdom remains. It was one of the challenges that Rudolf Steiner posed to his listeners during his course on agriculture, namely to create a new and conscious form of intuitive farming wisdom. This would involve developing a form of 'clair-sentience', a conscious sensitivity for life that enables an intimate communion with nature to take place. This communion, together with the daily engagement with the land, can give birth to new intuitive perceptions – like the farmer's 'knowing eye'. Such a deep connection to the land and the living spirit of the place is a quality that is found in indigenous cultures all over the world.

Once I set out to prepare a biographical story of the biodynamic movement it soon became clear that not everything could be included. Thus, fairly early on it became apparent that my focus would have to be limited to the evolution of biodynamics in Great Britain. It is, however, impossible to leave out the tragic split that occurred in the anthroposophical movement at the Goetheanum during the 1930s as this had reverberations that continued to be felt in Britain (and elsewhere) long afterwards. It is nevertheless remarkable that, unlike many other spiritual movements that split after the death of their founder, the two associations eventually came together again as one organisation. Today the world movement is more or less united.

Steiner's original Agriculture Course was given in central Europe to a German-speaking audience who were deeply embedded in the unique folk traditions of their landscape. On the British Isles, with their equally unique traditions and with an English-speaking culture, this new approach to agriculture inevitably developed in a different way. In Germany, most of the well-established biodynamic farms had been family farms for generations, while in this country many were started by 'incomers' to agriculture. This is undoubtedly connected with the industrial revolution and the associated earlier loss of peasant farming culture in Britain. Another aspect is that the aristocracy of central Europe appears to have been bonded more strongly to the land and the culture rising from it than that of Britain. Here the focus was more individualised and linked to national identity.

The biographical sketches I give of personalities in the British biodynamic movement are by no means exhaustive and there are many others who might have been included. However, while it has not been possible to describe everyone who has contributed to the movement's development, the attempt has been made to include the names of all those who became actively involved as Council members of the Anthroposophical Agricultural Foundation, the Biodynamic Association and, after the Second World War, the Biodynamic Agricultural Association. Tables with this information for each of the above organisations can be found in the appendices at the back of the book.

1. The Beginning

The story of the biodynamic movement has many layers. It begins in a small Silesian village in what is now Poland, some twelve miles south west of Wrocław (formerly Breslau). Koberwitz, or Kobierzyce as it is now called, lies on the open and largely treeless plains of the River Oder. Its 'black-earth' soil – what the Russians call *chernozem* – makes it one of Europe's most fertile agricultural regions and, since it is relatively flat and easily accessible, also one of the most exploited. Over the last few decades many of the large global agri-industrial corporations have set up their production units on the fringes of Wrocław.

This development, however, is nothing new. In Steiner's day technology was already coming to the large farming estates of lower Silesia along with newly developed artificial fertilisers. Sugar-beet production had long been a major source of income and it was now being grown on a scale that merited the building of railway lines to carry harvested roots from the fields to the coal-powered factories. Silesian coal mines were an important resource that supplied the energy for Germany's rapid industrial revolution. This fast rate of modernisation was possible because much of the land still belonged to the aristocratic families whose huge estates covered many thousands of acres.

The 18,500 acre (7,500 hectare) estate around Koberwitz where the Agriculture Course took place belonged at the time to Count

Carl Wilhelm von Keyserlingk. He asked Rudolf Steiner to share his insights into agriculture and managed, after a great deal of effort, to persuade him to come to Schloss Koberwitz. The story of Count Keyserlingk and the conference that took place at his home during Whitsun of 1924 sets the scene for the unfolding story of biodynamic agriculture in both central Europe and in the English-speaking world.

Count Keyserlingk was born on August 14, 1869 in Schloss Jakobsdorf near Wengeln in lower Silesia. His father, Eugen Wilhelm von Keyserlingk, was a widely travelled scientist, known for his research into spiders.[1] The family traces its ancestry back 500 years to the Baltic-German Chivalric Order. As well as owning large tracts of land stretching from Estonia through to Silesia, the family played a leading cultural role during the eighteenth and nineteenth centuries, with several family members becoming well known in the fields of philosophy and science. His mother, Margarete, was the daughter of a historian who served as the Bavarian ambassador to Rome. Carl grew up with his older brother, Robert, on the family estate but also spent some time in Rome as a child. After school he trained as an agricultural engineer and then spent some years as an army officer before returning to agriculture. In 1899 he married Johanna Skene, who came from a Scottish aristocratic family in Skene, not far from Aberdeen. They had three sons together, but the first one died soon after he was born.

Johanna von Keyserlingk spent most of her childhood in Silesia in the city of Breslau along with her two brothers. Her father had emigrated from Scotland while she was still small and had established a successful industrial-scale sugar-beet business. Her mother, Clara Schoeller, came from the German Rhineland. Johanna inherited a unique clairvoyant capacity, which Rudolf Steiner referred to on one occasion by saying that her Scottish inheritance made it entirely free of Luciferic influences. It was, he said, a form of clairvoyance that would become more widespread and normal as the third millennium progressed. Her unusual capacities become apparent to anyone who reads her contributions in *The Birth of a New Agriculture.*[2]

Figure 1.1: Count Carl Wilhelm von Keyserlingk.

Figure 1.2: Johanna von Keyserlingk.

Count Keyserlingk and Rudolf Steiner met for the first time shortly after the First World War. They discussed the problems associated with the increasing use of materialistic techniques in agriculture and the harmful effect they were having on the health of crops, livestock and humans. Steiner had at that time just set up an anthroposophical business initiative in Württemberg called Der Kommende Tag. The company was established as an associative economic partnership comprising around nineteen businesses and a research institute, a Waldorf school and a clinic in Stuttgart as well as a farming enterprise. By setting up a structure within which economic and cultural organisations can mutually support one another, it sought to demonstrate how aspects of the Threefold Social Order could be implemented practically. Recognising the Count's experience in this field, Steiner asked him to oversee the project.

The Count's main work, however, was with the firm run by his father-in-law, the sugar-beet processing company Rath Schöller & Skene AG based in Silesia. The company was at the cutting edge of the new scientific revolution associated with the growing chemical industry. A very moving account is given by Johanna von Keyserlingk of the struggles her husband endured as he sought to fulfil his obligations to his father-in-law's sugar-beet processing company, while at the same time helping Rudolf Steiner to develop Der Kommende Tag. She describes how, as soon as Carl took on these new responsibilities, life in the sugar factory took an unpleasant turn. Her father objected strongly to his activity and from that moment on sought to remove Carl from his leading position in the company. In hindsight it was a classic clash between an industrial company pursuing a materialistic quest for profit, and a new social and spiritual endeavour. It all came to a head with Carl being sent a letter of dismissal. Johanna von Keyserlingk describes how he showed this letter to Steiner:

> When Rudolf Steiner had read the letter in his usual careful way, he remarked: 'Well, the letter is not exactly polite, but you should go back nevertheless!' My husband was very surprised, because this piece of advice was actually an impossibility. Yet Carl accepted it with inner confidence.

She then goes on to describe how he went to his offices and found them cleared of everything:

> In this situation Carl obeyed the voice that had given him advice. He travelled by train next morning to Klettendorf, walked to the factory and sat in his office. No one greeted him, nobody brought him letters or called at the door. The employees seemed to have been given orders to this effect. When it was time to leave he went back home. The same thing happened the next day, and on the third day it was repeated once more. But then suddenly there was a knock at the door. The head of the factory stood there, white in the face and stuttering: 'Your honour! A deputation of the workers has arrived – a general strike has been declared at the Trades Union headquarters in Breslau if you do not immediately take up your post again!' Carl was very moved by what had taken place between him and Rudolf Steiner. The results of this situation exceeded all expectations. Not only was his authority restored to him, but he was given a new contract with complete freedom to work as he thought best and a higher wage. The unlimited power of my father had come to an end.[3]

Carl was also granted his long-held wish to take up residency in Koberwitz. The family moved there in 1920, four years before the course was given, and stayed until 1928, four years after the course had ended. Carl was a man whose life's purpose became the Agriculture Course and its realisation. A very special place and time was needed for Rudolf Steiner to deliver the course and without the enthusiasm and dedication of Carl and Johanna it would hardly have been possible. This was described by Johanna as the happiest period in their lives: four years of preparation followed by four years of grace.

Figure 1.3: Koberwitz House where the Agriculture Course was held in 1924.

Figure 1.4: Koberwitz House today.

By 1928 the former difficulties had returned and, due to the continuing opposition from his father-in-law's company, they were forced to sell up and leave. The Count then bought another estate where he planned to carry out a full conversion of the land to biodynamic methods. This would create a base upon which to undertake further research for the Experimental Circle, which he had helped to found during the Agriculture Course.[4] Despite his best endeavours, however, the Count found himself in conflict not only with the family firm but also increasingly with other members of the Experimental Circle. This eventually led him to resign from the leading role he held in the circle. With this step it seemed his life's task had been taken from him. On December 29, 1928, while on a journey to the Goetheanum, the Count died suddenly. He was fifty-eight.

Much has been written about the events surrounding the course in Koberwitz. Far from being simply another series of lectures, however, it was an event of great significance for the future of Europe and for the world. As has been mentioned, the timing was significant, as was the fact that Count Keyserlingk had a dedication to the land that had been cultivated by his family over many generations. The warmth of this inherited devotion and connection to farming life could now be placed in the service of a new impulse for agriculture – a new birth that was embedded in the warmth of a fast disappearing yet rich and integrated culture of the land. Gertrude Mier, who worked for a number of years in the Keyserlingk household in Koberwitz, describes the wonderfully social quality of the life there – the charm and vivacity of Johanna and the often silent yet immensely upright presence of the Count, a presence that exuded a much-loved and respected authority:

> When he was away on a journey, immediately difficulties arose, problems cropped up, people quarrelled – and one

> could almost hear the sigh of relief when he returned. Without any outward action or word spoken by him, harmony and peace returned.[5]

She describes how everyone was puzzled that his mere presence could have such an influence. Commenting on this Steiner said with a smile, 'People who were present at the mystery of Golgotha, can be truly unselfish.'

Reflecting on the time during which the Count and Countess resided in Koberwitz – the four years of preparation, the event itself, and then the four years that followed – a picture begins to emerge that seems to suggests a link to the chivalric past of the British Isles. Legend tells how, after the quest for the Holy Grail had been achieved and his Round Table began to disintegrate, King Arthur observed to his knights that their mission was now accomplished. Their moral tasks of knighthood had been fulfilled through the successful quest for the Holy Grail. Their deeds had shone like a bright light in the darkness of that time. Before the Round Table was founded there was darkness and after it dissolved there was again darkness. During their time at Koberwitz, in the four years leading up to the course and the four years after it, the Keyserlingks enabled the community light of Whitsun to shine from that great festival. Before their arrival it was dark and after they left it was dark, too. Their time at Koberwitz had been, like King Arthur's Round Table, a beacon of light between periods of darkness.

Many years later Carl Mier,[6] his one-time secretary, spoke of Count Keyserlingk as a knight and a fighter for truth, whose moment of destiny came when he met Rudolf Steiner and was able to prepare the setting for the Agriculture Course. He describes, too, the Count's remarkable hands, which were used to play the violin. After the

Count's death a cast was made of his hand with the forefinger pointing towards the earth. This cast still exists along with his death mask. They were given into the care of the Biodynamic Association in Britain. Though he never travelled to Great Britain, the Count felt increasingly drawn to the British Isles, and during the last years of his life prepared himself for one day coming here by learning to speak English. In the end it was Carl Mier, his scientific secretary, who made the journey and carried the impulse of biodynamics to these islands.

2. The Agriculture Course and the Experimental Circle

The Agriculture Course was given at Whitsun at the beginning of June 1924. It was experienced as a great Whitsun festival event, a celebration with an atmosphere enhanced by the grand setting of the Koberwitz estate and the hospitality of the Keyserlingk household. According to those present, when Rudolf Steiner arrived he was weak and clearly unwell, but as the course went on his health improved to such an extent that by the end he became almost youthful. There were 130 participants, most of whom were farmers, land workers or estate owners. Many of them afterwards admitted that although they were inspired and felt the rightness of what Steiner had said, they could not understand it intellectually. What the course brought was an entirely new way of imagining nature and of conceiving the farm. Fundamental aspects of farming, such as the improvement of soil fertility and plant health, were placed in the context of the totality of life and its interconnected relationship to the living forces of the cosmos.

After the first lecture on the Saturday there was a break of two days before the remaining seven lectures were given each day from Tuesday through to Monday. On Whit Sunday, Steiner took part in a walk around the Koberwitz estate and was able to experience what was living there: the iron-rich soil, the growing conditions, the climate, and much else. This enabled him in the days that followed to speak of

Figure 2.1: Attendees of the Agriculture Course.

agriculture out of the specific context of place, and in a very concrete and immediate way to introduce the fundamental concept of the farm as a self-contained individuality.

Practical aspects of the course focused on increasing the vitality of the soil by describing preparations that can be made to improve the manure and stimulate healthy plant growth; counteracting harmful influences and reducing pest damage; developing sensitivity for the intricate relationships in the web of nature between trees, insects and fungi, and providing insights into the nutrition of farm livestock and much more. A hundred years later these agriculture lectures continue to be a source of inspiration for generations of land workers.

The Experimental Circle

To explore these issues further Count Keyserlingk, together with the group of farmers who were present, came together with Rudolf Steiner during the conference to found the Experimental Circle. This was intended as a forum for sharing practical experiences and undertaking research into this new agriculture. The address given by Rudolf Steiner during the Koberwitz conference, considered by many as an integral part of the Agriculture Course itself, is the occasion when the Experimental Circle was formally introduced. In this address Steiner describes the context and purpose of the Experimental Circle in a very clear and direct way.

He points to the great and often undervalued wisdom of the traditional peasant farmer, whose lifestyle he had known intimately as a young man:

> I have always considered what the peasants and farmers thought about their things far wiser than what scientists were thinking … Far rather would I listen to the wisdom of someone speaking of experiences drawn from his daily work on the field than all the convincing Ahrimanic statistics arising from our highly developed science.[1]

This was not to undervalue the contribution made by rigorous scientific investigation, rather Steiner wanted to emphasise the importance of observing the phenomena for their own sake, whether in the laboratory or out in the field. He recognised how important it would be for the scientists undertaking research in the field of spiritual science at the Goetheanum to acknowledge the value of and develop a close working relationship to the practical wisdom living amongst farmers. This intuitive knowledge is no longer as strong as it once was and needs cultivating anew in a conscious and active way.

These considerations offer a key towards understanding what was really intended with the creation of the Experimental Circle. It was to be in the first place a forum where active biodynamic farmers and gardeners could meet and share their farming experiences, find solutions to the challenges confronting them in their work and address their inner questions and concerns. With regard to research it is often difficult for farmers, whose lives are devoted to enhancing the health and well-being of the land and livestock in their care, to consider setting up comparative trials. It is not only about finding time to do the extra work that this entails, it is also about the farmer's relationship to their farm. If out of experience or through intuition a farmer discovers a beneficial measure, should it not be applied immediately to the whole farm? Research on a farm often takes an intuitive form. A farmer will, for instance, look at an animal or a field of corn and as a result of years of experience and close observation somehow know what is needed. Living with and observing a phenomenon meditatively as well as directly, develops a capacity that can be described as having a 'knowing eye'. In former times this was an inborn capacity, today it is an ability that needs schooling. It is this intuitive quality that should live as a key aspect of the work carried out within the Experimental Circle.

The Experimental Circle of the Anthroposophical Society as it was first called (it later became the Experimental Circle of Anthroposophical Farmers) was in actual fact a community of farmers. It originally consisted of the practising farmers who attended the Koberwitz lectures; others then joined who were able to demonstrate a clear commitment to the cause. In the early days, great care was taken to protect the biodynamic method by holding it within a very limited circle of people. At its founding Rudolf Steiner laid down some fairly strict conditions for membership. He said:

> To begin with, only those members [of the Anthroposophical Society] should be admitted who (a) are in a position to carry out practical experiments in their own farms and gardens and in such a way that it is impossible for non-members of the Experimental Circle to ascertain how the preparations are made, or (b) are in a position to assist the development of the new methods by scientific researches or the like.[2]

This also meant that the agriculture lectures were not freely available but could only be circulated among members of the Experimental Circle. The early copies were in fact numbered and had to be returned when no longer needed. Steiner was concerned to ensure that the new approach should only come into the public domain once the methods had been scientifically and practically tested.

Count Keyserlingk felt particular responsibility for this and worked to ensure that strict adherence to these rules was maintained. His experience with the chemical companies and their desire to stop a real alternative from arising made him very cautious. He had spent many years working in his father-in-law's company and this had brought him into close contact with the major chemical and fertiliser company established in the area – IG Farben. It was perhaps no coincidence that the centre of its activity was very close to where the Count's first biodynamic trials were taking place. Sensing that the new approach might pose a threat to their business, the company took some interest in what was going on. It was not long after the Agriculture Course had been given that the Count was invited to meet with representatives of

IG Farben and offered a large sum of money in return for a transcript of the lectures. Keyserlingk turned this down and so they plied him with questions instead. He then noticed a strange dish on the table with a most unusual form – it was an early audio recording device. The shock of this discovery alerted him to the hidden motives behind their interest. It is probably also what lay behind his subsequent recommendation that each of the preparations should be referred to by a number: 500 and 501 for the spray preparations and 502 to 508 for the compost preparations and Equisetum tea.[3]

Count Keyserlingk was also concerned about a possible watering down of the spiritual and social aspects of the approach if too great a focus was placed on the outer applications. Not everyone agreed with him, of course, and as the work became more established it became increasingly difficult to maintain such a strict position, despite his strenuous efforts to do so. It was this more than anything else that ultimately led to him to relinquish his leading role within the Experimental Circle. It took many more years, however, before the inner core of biodynamic practice could come fully into the public domain.

In his address, Steiner emphasised how important it would be for the Experimental Circle to work closely with the School of Spiritual Science so that a new form of folk wisdom can 'fertilise a new living science'. [4]

The Experimental Circle founded during the Agriculture Course was an informal group until it gained legal status as an association in Germany on December 1, 1929. In Britain it was founded as an informal group in November 1928 and has remained so ever since.

Figure 2.2: Plaque commemorating the Agriculture Course given by Rudolf Steiner at Koberwitz during Pentecost, 1924.

The School of Spiritual Science and the Agriculture Section

Six months before the Experimental Circle came into being, the General Anthroposophical Society had been founded at the Goetheanum during the Holy Nights of 1923/24. One year before that, the first Goetheanum, an artistically crafted and utterly unique building that had been worked on throughout the war years and constructed entirely of wood, was set on fire and destroyed. This tragic event made Rudolf Steiner more determined than ever to continue with his work and so he set about creating not only a new building but a whole new organisation, a new anthroposophical society built in the image of the lost Goetheanum. Its primary objective was and is to nurture the life of soul on the basis of a true knowledge of the spiritual world. It was structured to be as open and accessible as possible: membership was to be for anyone with a positive and open mind who could see justification in the work of the Goetheanum.

Within this very public organisation Rudolf Steiner then founded and embedded the School of Spiritual Science as a vehicle for inner schooling and spiritual research into all the various vocations and disciplines of life. The Goetheanum was to become the campus of an independent University for Spiritual Science, differentiated into various sections or departments. Today the eleven sections include agriculture, education, medicine, natural science, astronomy and mathematics, visual arts, performing arts, humanities, social science as well as sections for general anthroposophy and young people. Until relatively recently agriculture was included within the Natural Science Section and only with the dawn of the new century did it develop its own identity and become a department in its own right. Although centred at the Goetheanum in Dornach, the School of Spiritual Science exists wherever people are actively engaged with it. This means accepting responsibility for the research that is undertaken and implies making a commitment to the school and fellow researchers, and an ongoing schooling in anthroposophy.

Although active spiritual investigation requires a degree of seership or clairvoyance, spiritual research also involves trying to understand and deepen the results of existing spiritual research, bringing personal experiences to bear on it and thoroughly penetrating the material produced by Rudolf Steiner in his many books and lecture cycles. These contain an enormous amount of material for deepening our understanding of life, of nature and of agriculture. Everything that is studied needs to be assessed, grappled with and, as far as possible, understood on its own terms and related to practical experience. To wrestle with these insights is part of spiritual research. Whoever has worked with Steiner's Agriculture Course for any length of time soon discovers that there is always something new to learn from it, no matter how many times they have read it. It is a tool for continually deepening our understanding of life, nature and how we work with the earth.

The Experimental Circle in the British Isles

The Experimental Circle of Great Britain and the English Speaking Countries came into being following the 1928 World Conference on Spiritual Science in London. Its purpose was to 'provide an informative background for the practical application of biodynamic methods'.[5] According to the veteran biodynamic farmer Alan Brockman:

> The early pre-war meetings of the Experimental Circle were very frequent, almost monthly it appears. This was possible since those involved were mostly 'gentlemen farmers' who could afford to devote the time and resources to do so. After the war things became more difficult and although the rift in the Anthroposophical Society gradually healed, the biodynamic work was reduced to a 'holding operation'.[6]

Regular meetings continued to take place at least once a year, however, and some fascinating and far reaching studies were undertaken. These explored many themes connected with the earth, agriculture and human evolution, and were based not just on the works of Rudolf Steiner but also on further insights contributed by members of the Circle. These studies served to enrich the daily practical work on biodynamic farms across the country and develop a social and spiritual understanding for the biodynamic 'mission'. Meetings of the Experimental Circle continue to this day. The esoteric link between Britain and Koberwitz has thus never been broken.

Years of esoteric work

In the immediate aftermath of the Second World War, a number of individuals decided to spend one day each month at Rudolf Steiner House studying background material to biodynamics. Among them were David Clement, Carl Mier, George Adams, Olive Whicher, Marna Pease and Doris Davy. Their intention was to re-energise the work for the post-war period and, in 1947, they held the first of the annual weekend conferences of the Experimental Circle. A number of visiting lecturers were invited to these events, including Ernst Lehrs, Herman Popplebaum, Rudolf Hauschka, Bernard Lievegoed and Karl König. The meetings were described by David Clement as being:

> warm, enthusiastic occasions to which about twenty members and invited guests came. Four lectures were given between Friday evening and Sunday midday and one session for questions and any business that had to be done.[7]

Conferences during the immediate post-war period were held at Peredur, a school established in East Grinstead, West Sussex, for children who had difficulty adapting to mainstream educational institutions. It remained the venue of choice until 1975 when Peredur relocated to Cornwall. According to Katherine Castelliz these conferences continued to be set up around invited speakers. It was only in the early 1970s that members of the Experimental Circle began to create their own content and work on it together. This approach brought renewed vigour to the meetings and it remains the preferred way of working to this day.[8]

Continuation and renewal

Following the split in the Society in the 1930s, links between the movement in Britain and the Goetheanum were often tenuous, a fact exacerbated by there being two completely separate associations between 1935 and 1950. Even after the war, when tireless efforts were being made to build bridges to the continent, the biodynamic movement in the British Isles continued to follow its own independent course to a large extent. It wasn't until many years later, nearly eighty years after the Experimental Circle had been founded in Britain, that a fundamental new step was taken.

At a meeting held at Hood Manor in Devon in February 2006, the decision was taken to link up with the newly formed Agriculture Section, which had recently emerged from the Natural Science Section, and thereby establish a UK branch of the Section working in the field of agriculture. This move brought to conscious expression the two aspects of work within the Experimental Circle, cultivating a deeper understanding of the biodynamic foundations and undertaking practice-based research, and led to the realisation of one of the original objectives set out during the Koberwitz conference, namely to achieve an ever-closer working together with the scientific wellspring of anthroposophy.

It was agreed at the meeting that:

- The Experimental Circle is a conscious expression of the Agriculture Section in the UK.
- Membership is open to *all* biodynamic farmers, gardeners and foresters who are also members of the Biodynamic Association (BDA).
- Membership is also open to any member of the School of Spiritual Science who would like to contribute to the work.

Annual meetings of the circle have continued to this day, supplemented from 2021 with monthly online meetings on various subjects and insights.

3. Daniel Dunlop and the Anthroposophical Agricultural Foundation

The initiative to bring biodynamic agriculture to Britain came from Daniel Dunlop, a British contemporary of Rudolf Steiner who was born in Kilmarnock, Scotland. His father was an architect and a committed, indeed zealous, Quaker. Daniel's mother died when he was only five years old, and he spent much of his childhood in the house of his maternal grandfather on the Isle of Arran, where he was surrounded by fisherfolk, ancient stones and daily Bible studies. His interests as a young man led him into practical engineering and the manufacture of machines. This work eventually led him to America where he became a well-respected industrial manager. He then took up an appointment as the European marketing director for a major electrical firm called Westinghouse Ltd. At the turn of the century he returned to Britain, settling in London with his family. Some ten years later, in 1911, he became the first director of the newly founded British Electrotechnical and Allied Manufacturer's Association (previously called the National Electrical Manufacturer's Association) and in 1924 chaired the first World Power Conference.

Throughout this period he studied the works of Helena Blavatsky, as well as other theosophical writers, and edited a journal for the Theosophical Society. He saw Rudolf Steiner for the first time in around 1906 but didn't meet him personally until 1922. Dunlop immediately recognised him to be an initiate while Steiner famously

Figure 3.1: Daniel Dunlop.

referred to him as his brother. From then on the two worked closely together albeit on either side of the English Channel.

It was largely thanks to Dunlop and his colleague Eleanor Merry that Steiner was able to give his great lecture cycle *The Evolution of Consciousness* in Penmaenmawr, Wales. This was a ground-breaking event on the fringes of western Europe that helped prepare for the founding of the School of Spiritual Science, which in turn complemented and paved the way for the Koberwitz conference the following year in the eastern part of Europe. Reflecting on his visit to the ancient stones of Penmaenmawr and the afterglow of the ancient druid culture, Steiner gave a series of lectures in Stuttgart about Druid initiation science. These lectures describe how in the ancient Druid culture a whole science existed that was based on reading the spiritual wisdom of the sun in the darkness of their great stone cromlechs or in the shadows cast by the stones. As the seasons progressed, the Druids observed how the light from the October sun had a different inner quality to that of July, and likewise how it had a different influence on the life of plants. They described elemental beings that grew to enormous size and manifested outwardly in the frost, the snow and the heat, and they taught how their forces as they worked in plants could be transformed into medicine. Such a deep understanding of the elements, along with the wisdom of the stars gained from the stone circles, provided guidance for agricultural tasks. These lectures consider many issues connected with agriculture that were not addressed in the Koberwitz course and therefore serve as a helpful companion to the agriculture lectures.[1]

Figure 3.2: Rudolf Steiner with the participants of the International Summer School held during August 1923 in Penmaenmawr, North Wales.

After Steiner's death, Dunlop dedicated himself to the dissemination of anthroposophy across the western world. He organised several big conferences, facilitated the translation of many key texts and arranged the first World Conference on Spiritual Science and its Practical Applications in London at the end of July, 1928. This event can be seen as marking the moment when anthroposophy was born into the English-speaking world.

For the conference Dunlop sought to bring together leading representatives of the practical fields of anthroposophy. These included Karl König, Ita Wegman, Eugen Kolisko, Friedrich Rittelmeyer and many others from across Europe as well as the UK. Amongst those invited was, of course, Count Keyserlingk as the representative of the farmers who had been inspired by the events in Koberwitz. Keyserlingk had long wanted to visit England but due to other commitments he had reluctantly to decline the invitation. In his place he sent his scientific advisor, Carl Alexander Mier (then called Mirbt, he later changed his surname to make it easier to pronounce in English).

Carl Mier was born in Marburg, Germany, in 1902, the youngest of five children. His father was a well-known theology lecturer at Göttingen University, and his mother was the daughter of a geography professor. On leaving school Carl studied agriculture in Göttingen and spent time working on farms in northern Germany before receiving his diploma in 1926. While still a student he met Count Keyserlingk in Koberwitz. He started working for him and soon became the Count's scientific advisor. In 1927, Carl married Gertrude, who was Keyserlingk's secretary at the Koberwitz estate. They had four children, one of whom died in childhood.

Figure 3.3: Carl and Gertrude Mier in 1954.

At the conference in London, Carl gave a talk appropriately titled 'Agricultural Depression: Its Cause and the Means for its Relief'. In it he gave a full account of what the new agriculture could offer. The talk was well received and it was after this conference that Daniel Dunlop invited Carl Mier to Britain where they founded the English-speaking branch of the Experimental Circle.

In order to create a legal vehicle to support the Experimental Circle, a further step was taken. In November 1928, just a few months after that seminal conference in London, the Anthroposophical Agricultural Foundation (AAF) was founded with Marna Pease, Maurice Wood, George Adams (then called Kaufmann, he would later change his surname), and Daniel Dunlop as its first executive committee, and with Carl Mier as its advisor. Dunlop retained a leadership role in the organisation until his death on Ascension Day, May 30, 1935.

This step of creating the AAF was felt necessary so that in the first instance a formal link could be made between the English-speaking world and the world headquarters at the Goetheanum in Dornach. A sum of money was donated to inaugurate the work, guarantee its first three years, and support Carl as a full-time consultant. When the three years came to an end and the initial funding ceased, a new structure was needed that would allow the AAF to access membership income. It therefore adopted a new constitution in 1931 and appointed Daniel Dunlop, who was at the time also General Secretary of the Anthroposophical Society in Great Britain, as its chair. Carl Mier became its secretary. A year later there were seventy-five members and four active biodynamic farms in the country. The need to increase membership meant that Carl Mier had to travel widely across the country to meet and inspire new members. A car was put at his disposal in 1932 for this purpose. Numbers gradually increased during the course of the next eight years and by 1940 the AAF had nearly three hundred members.

The AAF had four key objectives:

1. To serve the needs of members of the Experimental Circle. This meant in the first place assisting with the translation of lectures and other material.
2. To provide an organ of communication for both Circle members and those who were interested in finding out more through a circular called *Notes and Correspondence.*
3. To keep in touch with enquirers by letter or through a visit by the Biodynamic Advisor.
4. Finally, to address the growing interest in the quality of food and nutrition.

Figure 3.4: The journal Notes and Correspondence *published by the AAF.*

The role of the Experimental Circle in the context of the AAF was to provide information and support for the practical application of biodynamic methods.

From the time he arrived in England, Carl Mier dedicated himself to the biodynamic movement. In his capacity as Biodynamic Advisor, he travelled widely across the country meeting members and visiting farms. He also wrote regularly for *Notes and Correspondence*. In one of his pieces entitled 'Agriculture and the Present Age', published in the August edition of 1937, he discussed the inner mission of the farmer within both the difficult context of the time and the challenges presented by the new agriculture. He concluded with what could be described as a mission statement for biodynamic farming:

> What is it, that the spirit of the time calls to the farmer?
> To realise the earth is a living organism and to treat it accordingly.
> To understand the farm as an individuality; to maintain it as a well-balanced organism.
> To be aware of our responsibilities towards soil, plants and animals and our fellow workers on the farm.
> To shoulder with joy and seriousness the great task of providing humanity with food which not only nourishes the body but can enable man to realise the spiritual life as well.
> To play an active part in the whole social and economic life of the nation and of the world.
> So to work that farming shall remain the noblest vocation of man, whose task it is to make this earth of ours the dwelling place of future generations, that these may find a living earth as the place where they in their turn can fulfil the tasks of their human incarnations.[2]

When he retired from this work in 1956, Carl and Gertrude were invited to Camphill by Karl König, where they were soon involved in developing the new agriculture-based community of Botton Village in North Yorkshire. Some years later they moved to Delrow community and in their latter years made frequent visits to Germany and Eastern Europe. It was on one of these journeys while visiting the Wernstein community in Germany that Carl Mier suddenly died on June 22, 1975.

❁

The AAF found a home in Berkshire at the Old Mill House in Bray-on-Thames, near the town of Maidenhead. It was a large house belonging to Marna Pease, the long-serving honorary secretary to the Foundation.

Marna Pease discovered anthroposophy in 1920, but it was her experience of the 1923 Penmaenmawr summer school, where she also met Rudolf Steiner, that inspired her to dedicate her life to the cause. When she heard about the new biodynamic approach to agriculture she immediately set about applying it to her farm in Northumberland, which she managed with her husband, Howard Pease, a successful antiquarian from Middlesbrough. Otterburn farm was a beautiful estate with an imposing castle and some 1,280 acres (518 hectares) of land dedicated to stock rearing. Marna then offered to run the new Foundation, which she initially did from her home on the farm. When she took on the secretaryship in 1934 she was already well into her sixties. A year or two later, however, her husband died, and she was left to manage the farm on her own. She found this very demanding and also felt very isolated. It was Dunlop who then suggested that she give up the farm and move down south to the house that she had bought near Maidenhead with her half-sister, Eleanor Merry. Marna was at first very reluctant to leave the farm on which she had spent so much of her life, but finally she agreed.

Marna's new home, the Old Mill House in Bray-on-Thames, soon became the headquarters of the AAF. It was the place where the biodynamic preparations were made and distributed, and where Carl and Gertrude Mier lived with their young family during their first years in England. The house became a regular meeting place and was fondly referred to as the Anthroposophical Country Centre. It was used for small gatherings and as a visitor centre by both the Foundation and

Figure 3.5: Marna Pease.[3]

the Anthroposophical Society. In 1938 a small laboratory space was established there so that Lili Kolisko could carry on her research after she had to leave her research institute in Stuttgart.

The garden of the Old Mill House in Bray became a show piece of biodynamic gardening. The enthusiasm of Marna Pease and the tireless work of her gardeners transformed that little piece of Berkshire into a biodynamic oasis. It was one of the first – if not the first – biodynamic garden in the country. In an article for *Notes and Correspondences*, Marna's friend and long-time colleague, Olive Mainland, described the garden's development, the many unusual (at that time) vegetables growing there, and the bee garden:

> A great work was the transformation of a hard tennis court into a most beautiful garden growing special plants for her bees. The gardeners dug deep and to Mrs Pease's design, made lovely sweeping beds for the plants and put a little bee fountain in the centre with a shallow bowl where the

> bees could alight and drink. The bee garden was enclosed by wattle hurdles and the gaily painted hives stood in rows with a lovely background of vines and purple clematis. A little low stool was kept in this garden and when a swarm was expected Mrs Pease sat for hours ready to deal with it. It was most impressive to watch her skill. She always had a syringe with her to make a gentle rain.[4]

The grounds extended down to the riverbank and the property included the old mill room, an ancient timber-framed building mentioned in the Domesday Book. This is where honey was extracted and herbs dried. Olive went on to describe how jars of pot-pourri were stored there, often for several years and growing sweeter as time went by. The biodynamic preparations were made there, too, and stored in a little room above the mill stream before being packed and mailed out to biodynamic gardeners across the country. Marna was much loved by all who met her. She had a tremendous capacity for warmth and was interested in everyone. According to Olive those with no connection to biodynamics would observe, 'Mrs Pease is such a dear, but what a pity she has such a bee in her bonnet about the planets!'[5]

The Old Mill House continued to be the home of the AAF with Marna Pease working tirelessly as its secretary right up until the end of the Second World War when she retired. She died following a stroke, aged eighty-one, on August 31, 1947. The Old Mill House still exists today as a private house with extensive gardens.

Figure 3.6: The Old Mill House in Bray-on-Thames.

Sleights Farm: the first biodynamic farm

The heart of any agricultural movement lies with those who work the earth as practical farmers. The person who can be described as the first biodynamic farmer in Britain was Maurice Wood of Sleights Farm in Yorkshire. Maurice Wood first heard about the new agriculture when he attended the World Conference for Spiritual Science and its Practical Applications in London in 1928. As a result of hearing Carl Mier's lecture, Maurice, who was already an active farmer, decided to convert his farm in Yorkshire and pioneer practical biodynamic work in Britain.

Maurice Wood was born in 1884 in the industrial city of Leeds. He was the son of a successful builder and, reaching further back, his forefathers had been tradesmen and merchants. When Maurice finished secondary school he went to college and studied business management. When he graduated he quite naturally took up his father's vocation and established himself as a builder. Many of the houses in Leeds as well as other buildings in the city were constructed by him, and a fair number of these are still in existence today. During this time Maurice met his wife, Etta. They settled down and soon started a family. In 1910 their daughter, Janet, was born and in 1913 her brother, John.

This peaceful and secure existence came to an abrupt end in 1914 with the outbreak of the First World War. At the time, Maurice was thirty years old with a promising career in front of him, but nothing would be the same again. In the aftermath of the war, his efforts to pick up the pieces of his life in a world so utterly different caused him to think long and deep about the future. Should he return to the business world and pick up where he left off or should he begin something far more creative? His friends and relatives all advised Maurice to return to the work for which he had been trained. After all, why should the opportunity of a good career be discarded?

Figure 3.7: Maurice Wood.

To do anything else would be uncertain. However, Maurice knew that his former business had not been able to meet his family's needs – when it came to the crunch it had been unable to provide the basic necessities of life. He became convinced that the only sure way to rebuild their lives was to grow their own food. Everything else would follow on from that. And so this city boy embarked on a journey that was to lead to a lifelong commitment to agriculture.

Having come to this decision Maurice immediately set out to find suitable land. It wasn't long before he found and bought a farm about twelve miles to the north of Leeds outside the village of Huby. Sleights Farm was a 15-acre (6-hectare) holding set in rolling and well-wooded landscape on the edge of the moors some 500 feet (150 metres) above sea level. The buildings and farmhouse were more or less derelict and the land, though fairly fertile, was pretty neglected. Despite the enormous amount of work facing them, he moved there with his wife and two young children in the autumn of 1919. They had much to get used to: the silence of the countryside, the darkness of the night away from the city lights, and the knowledge that from now on they would have to make their living from the soil, not to mention carry out the renovations required for their dwelling.

The main focus of the farm enterprise consisted of hens (since Maurice figured that eggs would always be in high demand) and vegetable production. From the very beginning he carried the ideal of a self-sufficient farm and planned to feed his birds on what the land could provide. Not being a born farmer – an uncommon situation in those days – he knew that he had much to learn if he was to be successful in his new profession. He always kept his eyes and ears open for any hints and good advice that might come his way. Maurice once explained that an old hedger he employed in the early years of the enterprise became a priceless source of rural knowledge and agricultural wisdom. Under his guidance, Maurice grew more confident as a farmer and gradually managed to support his family from the land.

Seven years after taking on the farm, Maurice suffered from a severe attack of shingles that knocked him out of action for many weeks. While he was recuperating something happened that was to change his life. He was invited by a close friend to attend a conference on the Isle of Anglesey off the north-west coast of Wales. It was here in 1926 that Maurice Wood met George Adams. Born of a German father and an English mother, George's mastery of both languages had gained him wide respect as an interpreter and he had been Rudolf Steiner's translator during his lecture tours of Britain. Not only was he completely at home in English and German, he also had a remarkable memory. When Rudolf Steiner gave a lecture (which he always did in German), he broke it into three parts. After he had spoken for some twenty minutes, George Adams would repeat the entirety of what he had said in English – a scarcely imaginable feat of memory, concentration and understanding, especially with a subject that was so new and challenging. Along with Maurice, George Adams would later become one of the four founders of the AAF.

Shortly after the conference in Wales, Maurice joined the Anthroposophical Society. Two years later he attended the London conference and, inspired by what he heard, set about making his farm the first biodynamic farm in Britain.

During the conversion it became clear that many things had to change, not least the reliance on poultry, which at that time included 1,000 laying hens, 400 ducklings and 80 geese. There was no way he could provide feed for so many of them from the farm. Therefore, in 1930, he decided to sell off most of his poultry and, in response to many requests for biodynamic corn and especially for wheat, he began

Figure 3.8: George Adams.

cereal production instead. He rented a further 23 acres (9 hectares) of moderately good land and started to grow wheat.

This step soon necessitated a further one and, in 1939, Maurice designed and installed a purpose-built flour mill in order to distribute fresh, stone-ground flour to his customers. Sleights Farm quickly became known across the country for its freshly milled biodynamic flour. The mill design proved such a success that several more were constructed. Some of these are still in use today on other biodynamic farms. Maurice's primary concern in designing the mill had been to retain the highest nutritional quality. He sought to achieve this by avoiding the production of excessive heat during milling and by remaining as close to operating a natural process as possible. His design therefore sought to ensure that milling proceeded no faster than the speed at which a horse can chew oats!

Maurice became a regular writer for *Notes and Correspondence.*

Through his writings he was able to share his understanding of biodynamic husbandry with a wider audience. His articles addressed issues such as how to develop a farm organism, working with the biodynamic preparations, understanding the effects of silica, and coping with the immense economic difficulties of the time. They remain to this day of great interest and are well worth reading.

Sleights Farm's central position in the country made it an accessible venue for meetings and conferences of the Anthroposophical Society and many of its leading personalities made regular visits to the farm. It was here that George Adams, assisted by Maurice Wood and Olive Whicher, translated the Agriculture Course into English.

Those who knew Maurice remember his down-to-earth and practical outlook on life. Short and somewhat stocky, he came across as a dour and earthy Yorkshireman. His dry sense of humour and tremendous warmth of heart, however, endeared him to all who met him. He was a keen geologist and collected stones and rocks from every place he visited. This occasionally proved too much for his ever-patient wife who would then remove them outdoors with the comment that since they had withstood so many aeons of weathering, a few more years would do them no harm.

The farm continued operating throughout the Second World War when its value as a food producing resource was widely acknowledged. After the war, circumstances were fundamentally different. The rapid industrialisation of agriculture and the demands placed on the land had changed so much that a completely new start was required. Maurice Wood continued for a further ten years until in 1956 at the age of seventy-two he decided to sell the farm and retire from active farming. It must have been an extremely difficult decision for him to make, but it is worth remembering that Sleights Farm had operated as a biodynamic farm for a full Saturn cycle of twenty-eight years.

4. The Challenge of the 1930s

The 1930s was a tumultuous decade for Europe. It began with the Great Depression and the rise of fascism and ended with the Second World War. It was also a very active though difficult decade for the Anthroposophical Society. In the years after Steiner's death in 1925, differences between leading members of the Society became ever more apparent, and without the spiritual leadership of its founder, relationships started to break down. By 1932 it was clear that a split in the Society was imminent and this was formally confirmed at the AGM in the spring of 1935. Two close colleagues of Rudolf Steiner – Ita Wegman and Elisabeth Vreede[1] – were voted off the Executive Council and six leading members, including Eugen Kolisko, George Adams, Karl König and Daniel Dunlop, had their membership of the Anthroposophical Society revoked. The Anthroposophical Society in Great Britain was also expelled. This tragic split meant that the anthroposophical movement was in turmoil at a crucial time in world history: the National Socialists had taken power and Europe was once again heading towards conflict.

This sad state of affairs had a knock-on effect on the wider movement. In Britain, the English Section of the Anthroposophical Society (allied to the remaining Goetheanum Executive) broke away from the Anthroposophical Society in Great Britain and formed a direct link to the Goetheanum. The split affected the entire movement and, encouraged by the English Section and the leadership at the Goetheanum, a new association for biodynamic agriculture was founded in 1933. Several members of the council of the Anthroposophical Agricultural Foundation resigned and in October 1937 joined the newly formed Biodynamic Association (BDA). For twelve years the BDA and the original AAF lived parallel lives with little communication between them. The consequences of this split continued to reverberate for many years.

The Biodynamic Association for Soil and Crop Improvement (BDA)

The Biodynamic Association had its seat at the Priory in Kings Langley, Hertfordshire. Margaret Cross and Leslie L. Binnie, both of whom had been previously active in the AAF, formed the new association with Barbara Saunders-Davies and Dr Hermann Poppelbaum, whose experience with medicine and Goethean science provided a good basis for promoting biodynamic farming and gardening. Lady McKinnon, an enthusiastic biodynamic gardener, also joined and chaired the organisation for several years. The organisation had a president and two vice-presidents. The first president was Lady Merthyr, well known for her work with the Girl Guide movement in Wales. Her two vice presidents were Harry Collison, General Secretary of the Anthroposophical Society in Great Britain until 1927, and F. Ferguson. By 1939 membership had grown to 111. The organisation produced the biannual *News Sheet*, which published essays and reports from farmers and growers in the biodynamic movement across the world. It was Margaret Cross in her capacity as Secretary to the BDA who produced the *News Sheet* and who remained its editor from its inception in 1935 right through to 1951.

Margaret Cross was born in Preston. Her mother died when she was two years old and she was cared for by her grandmother until she, too, died. Margaret's later childhood was spent with her father's family on their farm in Cambridgeshire. She studied education at Cambridge University and there met her life-long friend and colleague, Hannah Clark. Together they set out to establish what was to be possibly the first co-educational school in the country. In 1909 they took on what became the Priory School in Kings Langley. Margaret Cross became its headmistress and led the search to find new and innovative approaches to education. In 1922 she was invited to attend a course of lectures on education given by Rudolf Steiner. She knew then the kind of education she wanted to see develop in Britain and sought to include what she had learnt within the Priory School curriculum. It was only later, in 1949, that the school became a full Steiner-Waldorf school. Key to Margaret's approach throughout these years was her insistence that pupils in the school should participate actively and practically with cooking meals, cleaning windows, caring for the poultry, cattle and bees, and growing vegetables on the land at the Priory. Following his visit to the school, Rudolf Steiner gave a very positive report about this approach to his colleagues in Dornach and commented that: 'A child is thus led into life in a many-sided way and learns a whole mass of things.'[2] Biodynamic agriculture was of great interest to Margaret Cross. In 1929 she joined the newly established Experimental Circle and, soon after, the council of the AAF.

Margaret Cross remained with the BDA until it was re-united with the AAF in 1951. She died on March 31, 1962.

5. The Early Discoveries

The astronomical calendar

Elisabeth Vreede, head of the Section for Mathematics and Astronomy at the Goetheanum from 1924 to 1935, produced a calendar for farmers and gardeners with all the aspects and movements of the planets. She began producing this annual calendar in 1929 and it was translated into English and published by the AAF. With Elisabeth Vreede's abrupt departure from the Executive Council in 1935, the calendar ceased being published by the Goetheanum. The AAF then decided to publish an English version with Elisabeth Vreede's blessing. This new edition was published for the first time in 1937/38 and ran from Easter to Easter. It took the form of a handy, hardback, pocket-sized diary. It was packed with astronomical information but had no specific recommendations for gardeners. The calendar provided a weekly overview of planetary aspects and then gave details of specific astronomical events on the day they occurred. The position of the moon was given each day along with its rising and setting times. The times of sunrise and sunset were given twice a week and there was a well written explanation of the calendar and all its symbols by Elisabeth Vreede, as well as a table to show the times when each planet was visible.

Figure 5.1: Elisabeth Vreede.

Figures 5.2a–d: Sample pages from the 1937/38 calendar for farmers published by the AAF, based on Elisabeth Vreede's calendar.

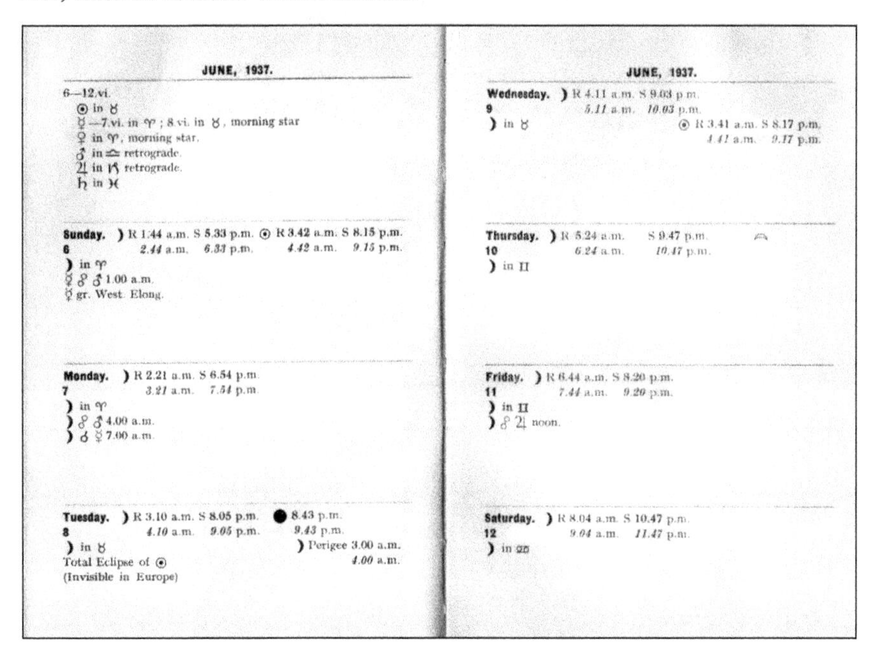

JUNE, 1937.

6—12.vi.
☉ in ♉
☿—7.vi. in ♈ ; 8.vi. in ♉, morning star
♀ in ♈, morning star.
♂ in ♎ retrograde.
♃ in ♑ retrograde.
♄ in ♓

Sunday. ☽ R 1.44 a.m. S 5.33 p.m. ☉ R 3.42 a.m. S 8.15 p.m.
6 *2.44* a.m. *6.33* p.m. *4.42* a.m. *9.15* p.m.
☽ in ♈
☿ ☍ ♂ 1.00 a.m.
☿ gr. West Elong.

Monday. ☽ R 2.21 a.m. S 6.54 p.m.
7 *3.21* a.m. *7.54* p.m.
☽ in ♈
☽ ☍ ♂ 4.00 a.m.
☽ ☌ ☿ 7.00 a.m.

Tuesday. ☽ R 3.10 a.m. S 8.05 p.m. ● 8.43 p.m.
8 *4.10* a.m. *9.05* p.m. *9.43* p.m.
☽ in ♉ ☽ Perigee 3.00 a.m.
Total Eclipse of ☉ *4.00* a.m.
(Invisible in Europe)

JUNE, 1937.

Wednesday. ☽ R 4.11 a.m. S 9.03 p.m.
9 *5.11* a.m. *10.03* p.m.
☽ in ♉ ☉ R 3.41 a.m. S 8.17 p.m.
4.41 a.m. *9.17* p.m.

Thursday. ☽ R 5.24 a.m. S 9.47 p.m.
10 *6.24* a.m. *10.47* p.m.
☽ in ♊

Friday. ☽ R 6.44 a.m. S 8.20 p.m.
11 *7.44* a.m. *9.20* p.m.
☽ in ♊
☽ ☍ ♃ noon.

Saturday. ☽ R 8.04 a.m. S 10.47 p.m.
12 *9.04* a.m. *11.47* p.m.
☽ in ♋

JUNE, 1937.

HISTORICAL NOTES.

1. 1826, The Pastor Oberlin of Alsace d.
2. 597, Baptism of King Ethelbert (Christianity in England). 1882, Garibaldi, Italian patriot, d.
3. 1657, William Harvey, discoverer of the laws of the circulation of the blood, d.
5. 1723, Adam Smith, English economist, b. 1906, Eduard von Hartmann, German philosopher, d.
6. 331, Julian, the Apostate, b.
8. 632, Mohammed, the prophet, d.
9. 597, St. Columba d. 1781, George Stephenson, perfector of locomotive, b. 1870, Charles Dickens d.
13. 323 B.C., Alexander the Great d.
15. 1215, Magna Charta (Birth of English liberty).
16. 1361, Johann Tauler, mystic, d.
17. 1703, John Wesley b.
18. 1916, Hellmuth von Moltke d.
19. 325, Council of Nicea.—1623, Pascal b.
22. 1527, Macchiavelli b.—1767, W. v. Humboldt b.
23. 1447, Christopher Columbus, discoverer of America, b.
24. THE DAY OF ST. JOHN THE BAPTIST AND OF THE SUMMER SOLSTICE. 1314, Battle of Bannockburn.
26. 363, Julian, the Apostate, d.
28. 1712, J. J. Rousseau, b. 1914, Archduke Francis Ferdinand murdered.
29. 1577, P. P. Rubens, the painter, b.

JUNE, 1937.

VISIBILITY AND MOVEMENTS OF THE PLANETS IN THE ZODIAC.

MERCURY as morning star 6.vi. reaches its greatest western elongation from the Sun (24°), remaining however invisible in the bright light of dawn.
Movement : 1—7.vi. in ♈ ; 8—30.vi. in ♉

VENUS remains longer visible as morning star, though its brightness slowly decreases. 27.vi. Venus reaches its greatest western elongation from the Sun (45½°).
Movement : 1—30.vi. in ♈

MARS is retrograde, as in May, until 27.vi., reaching below the stars of Libra the sharp second turning point of its loop begun 14.iv. Mars can be well observed especially in the first half of the night in the southern sky.
Movement : 1—30.vi. in ♎

JUPITER appears in the middle of the month before midnight in the south-eastern sky, standing during the whole second half of the night in the south.
Movement : 1—24.vi. in ♑ ; 25—30.vi. in ♐

SATURN can be better and longer observed in the morning sky.
Movement : 1—30.vi. in ♓

THE YEARLY COURSE OF THE SUN THROUGH THE CONSTELLATIONS OF THE ZODIAC.

DATE.	POSITION OF THE SUN.
21.iii.	*Beginning of spring.* The Sun stands at the Vernal Point—*i.e.*, at the crossing point of the Sun's course (ecliptic) and equator, in the first stars of Pisces.
10—18.iv.	The Sun rising in Northerly direction above the equator crosses the zig-zag ribbon of Pisces.
24—27.iv.	Passing the two main stars of Aries.
20/21.v.	The Sun stands below the Pleiades.
1.vi.	The Sun stands above Aldebaran, the main star of Taurus.
15.vi.	The Sun stands between the opening horns of Taurus.
21.vi.	*Summer solstice.* The Sun reaches the highest point of its course between Taurus and Gemini.
27.vi.	The Sun reaches the first stars of Gemini.
15.vii.	The Sun stands below Castor and Pollux.
31.vii.	The Sun stands near the star cluster Præsepe in Cancer.
24.viii.	The Sun stands near Regulus in Leo.
24.ix.	*Autumnal Equinox.* The Sun in its descending course crosses the equator, in the beginning of Virgo.
27.x.	The Sun stands above Spica in Virgo.
8.xi.	The Sun stands near the main star of Libra.
26.xi.	The Sun stands between the first stars in Scorpio.
1/2.xii.	The Sun stands above Antares, the main star of Scorpio.
21.xii.	*Winter solstice.* The Sun stands at the lowest point of its course, between Scorpio and Sagittarius.
6.i.	The Sun stands in the last stars of Sagittarius.
25.i.	The Sun stands at the head and horn of Capricorn.
13.ii.	The Sun stands at the tail of Capricorn.
6.iii.	The Sun leaves Aquarius.

TABLE SHOWING VISIBILITY OF THE PLANETS 1937-38.

Month	Planet invisible in the dawn	Visible in the Morning Sky: Planet rises before the Sun (hours)					Planets visible throughout the night	Visible in the Evening Sky: Planet sets after the Sun (hours)					Planet invisible in the dusk
			1	2	3	4			4	3	2	1	
March 1937 ...	☿		♃			♂			♀			♄	
April	♄			♃		♂					MidApr ☿ ♀		
May		♀	♄		♃		♂						☿
June	☿		♀	♄		♃	♂						
July				♀	♄		♃		♂				☿
August					♀		♄ ♃			♂		☿	
September ...			End Sep ☿	♀			♄	♃		♂			
October ...	☿		♀				♄	♃	♂				
November ...			♀				♄	♂	♃				☿
December ...		♀						♄ ♂		♃	Mid Dec ☿		
January 1938...	♀		End Jan ☿					♄ ♂					♃
February ...	♃	☿							♂	♄			♀
March		♃							♂		EndMar ☿	♄ ♀	
April			♃							♂	☿ ♀		♄

From this table one can see, for each month, which planets are visible, whether they are in the morning or evening sky and by how many hours they rise before or set after the Sun. The calculations are made as for the middle of the month. One has also, in the distribution of the planets from right to left, a certain picture of their distribution in space in the night sky, in the direction from West to East.

Astronomical considerations for sowing and planting

One of the earliest references to sowing seeds with regard to moon rhythms was written by Maria Hachez in the November 1935 edition of *Notes and Correspondence*. There she described some of the research carried out by the Section for Mathematics and Astronomy at the Goetheanum from 1930 to 1935. In addition to the waxing and waning cycle of the moon, a much broader field of cosmic influences was investigated, in particular, how the moon's position in the zodiac can affect the quality of the plants sown at specific times. The results she described differ from those later discovered by Maria Thun in showing other, less easily explained, differences in character.

The experiments were carried out using carefully selected seed from a number of different plant species, including vetch, peas, sainfoin, marjoram, sunflower, wallflower and cereals. The most frequently used, however, were the seeds of white and black radish since they can be used all year round. Seeds were sown on long beds outdoors and in pots and trays indoors using exactly the same soil. Sowings were made every two or three days during the appropriate constellation and always at the same time of day. Seeds were sown throughout the moon cycle starting with Pisces and repeated throughout the year.

Careful observations were made of the varied growth patterns, height and form, and some photographs were taken. The considerable differences in taste were also recorded. Here is a summary of the findings she described for each of the sowing constellations:

Pisces – regular and harmonious growth
Aries – ? [no particular growth pattern observed]
Taurus – plants were large and luxuriant
Gemini – plants were rather small and thin
Cancer – plants were small and stunted
Leo – regular and harmonious growth, often with rough and heavy leaves
Virgo – plants were rather small and delicate, but well formed and healthy
Libra – strong growth with especially large and broad leaves
Scorpio – much less and more irregular growth
Sagittarius – plants grew vigorously to the height of flowering plants
Capricorn – plants were small and stunted
Aquarius – ? [no particular growth pattern observed]

As regards taste:

> Aries, Leo and Sagittarius were sweet
> Taurus and Libra were almost tasteless
> Scorpio were unpleasantly sharp, bitter and with an aftertaste

Maria Hachez then concluded that to have fresh, well-grown, tasty radishes that are not bitter the seeds should be sown when the moon is in Sagittarius. In moist soil, seeds sown in Pisces and Leo are very good, while Capricorn and Cancer and especially Scorpio should be avoided.

Figure 5.3: Illustration from research carried out by the Section for Mathematics and Astronomy at the Goetheanum showing results of sowing radishes when the moon is in a different sign of the zodiac.

In summing up she wrote:

> We begin to feel these interweaving activities of planets and constellations as a language that speaks to us of heavenly wisdom. We must learn to read this writing of the starry heavens even as we learn to read our earthly vowels and consonants.[1]

Franz Rulni, who was present at the course in Koberwitz, produced a handwritten sowing and planting calendar for farmers and gardeners between 1948 and 1979. This was used widely in Europe for many years and included recommendations that were based on the moon's changing position in the zodiac. It was this calendar that led Maria Thun to conduct her many thousands of experiments into the effects on plants of the moon's sidereal rhythm.

Another very important figure in the biodynamic movement was Lili Kolisko. Lili dedicated her life to researching the foundations of biodynamic agriculture and anthroposophical medicine, but only found partial recognition from the wider movement. She was born in Vienna in 1889 and grew up in poor and challenging conditions with two step-sisters. We know little of her early years except that her father, who worked as a print-setter, suffered from alcohol addiction. Lili was a bright child. She attended the local Gymnasium and graduated with the Abitur, the German equivalent of English 'A' Levels.

In 1914, Lili volunteered at a hospital in Vienna, where she worked in a medical laboratory and learnt how to culture bacteria and distinguish the different cell types under a microscope. It was there in 1915 that she met her future husband Eugen Kolisko who was then practising as a junior doctor. The two shared a deep interest in science. Eugen introduced her to the work of Rudolf Steiner and gave her *Knowledge of Higher Worlds* to read.[2] It made a deep impression on her. She read it in a single night, followed in quick succession by other books by Steiner that Eugen possessed. In May of that year Lili met Steiner for the first time and it wasn't long before she asked him whether an esoteric chemistry could be developed. He suggested further training and offered a few concrete research suggestions. He also recognised her ability to 'see' the etheric. And so began Lili's life-long journey of research into both medicine and agriculture.

In 1920, the Koliskos, now married and with a one-year-old daughter, moved to Stuttgart to the newly founded Waldorf School. There, Lili set up the Biological Research Institute at the Goetheanum in a tiny room allocated to her. This was where she carried out her research into the spleen and worked with Steiner on a remedy for foot and mouth disease, which became known as the coffee preparation.

Figure 5.4: Lili Kolisko with her daughter, Eugenie.

Figure 5.5: Dr Eugen Kolisko.

Part of this research work required her to find out what homeopathic potency should be used. Following an indication given by Steiner, she grew wheat plants and then watered the seedlings with different potencies of the preparation to determine the most effective potency. This meant carrying out thousands of such tests on various medicines.

This work also enabled her to assess and to demonstrate the effectiveness of the smallest entities.[3] It was a major step in anthroposophical research and opened the door towards a radically new understanding of matter or, as Rudolf Steiner said:

> Success was had in breaking down the purely material substance so that the astral/spiritual activity could appear. For if you split material substance into atoms as atomic scientists do, but in a way that brings out the activity of its functions, its forces, you are showing the good will, I would say, to permeate matter itself with spirit in order to cross over to where spirit is active.[4]

Lili's work stood in complete contrast with that of nuclear scientists whose work led ultimately to the creation and dropping of the atomic bombs at the end of the Second World War. She could see how important it was to balance this hugely destructive power with a true understanding of the interaction between earthly and cosmic forces.[5]

Lili spent many years researching the influences of the moon and planets and undertook very detailed experiments to ascertain the effect of various moon phases on plant growth. She began with wheat grains and then extended it to include a whole range of vegetables. Seeds carefully selected for quality and germinating capacity were sown in glass jars at full moon, new moon and at the waxing and waning quarters. There were 240 plant replicates (thirty seeds in eight jars).

This was repeated each month between October 1925 and January 1927. She made a number of interesting discoveries. Some were connected with the changing seasons and the amount of sunlight received, and others directly with the phases of the moon. Her overall conclusion was that seeds sown two days before full moon produced better plants and greater yields than those sown two days before new moon. Those at half-moon were intermediate. Furthermore, she found that root vegetables sown at full moon would never become woody while those at new moon were less likely to show symptoms of decay. This demonstrated a clear polarity between the forces of full and new moon.[6] Although Lili's research has been partly contradicted by the later work of Maria Thun, who placed greater emphasis on the moon's position in the zodiac and less on its phases, her attention to detail and thoroughness makes her findings eminently plausible.

It is Lili's perfection of the new technique of capillary dynamolysis for which she is best remembered, however. Using this technique, a solution of the plant juice or other substance to be tested is allowed to rise up a cylinder of filter paper. The paper is allowed to dry before being placed in a solution of metal salts, usually silver nitrate or gold chloride. As the solution rises up the paper, the patterns created by the plant juice become visible. The challenge is then learning to read the different forms and patterns that appear on the paper. After many trials Lili was able to read these patterns much as Ehrenfried Pfeiffer was able to do with his sensitive crystallisation pictures. Both approaches can open the door to gaining an image of the etheric formative forces that are active in the sap of plants.

During the 1930s the Koliskos found themselves embroiled in the conflicts that beset the Anthroposophical Society at the time. Lili Kolisko describes many of them in detail as part of a biography of her husband, which she published after the war. Feeling rejected by the

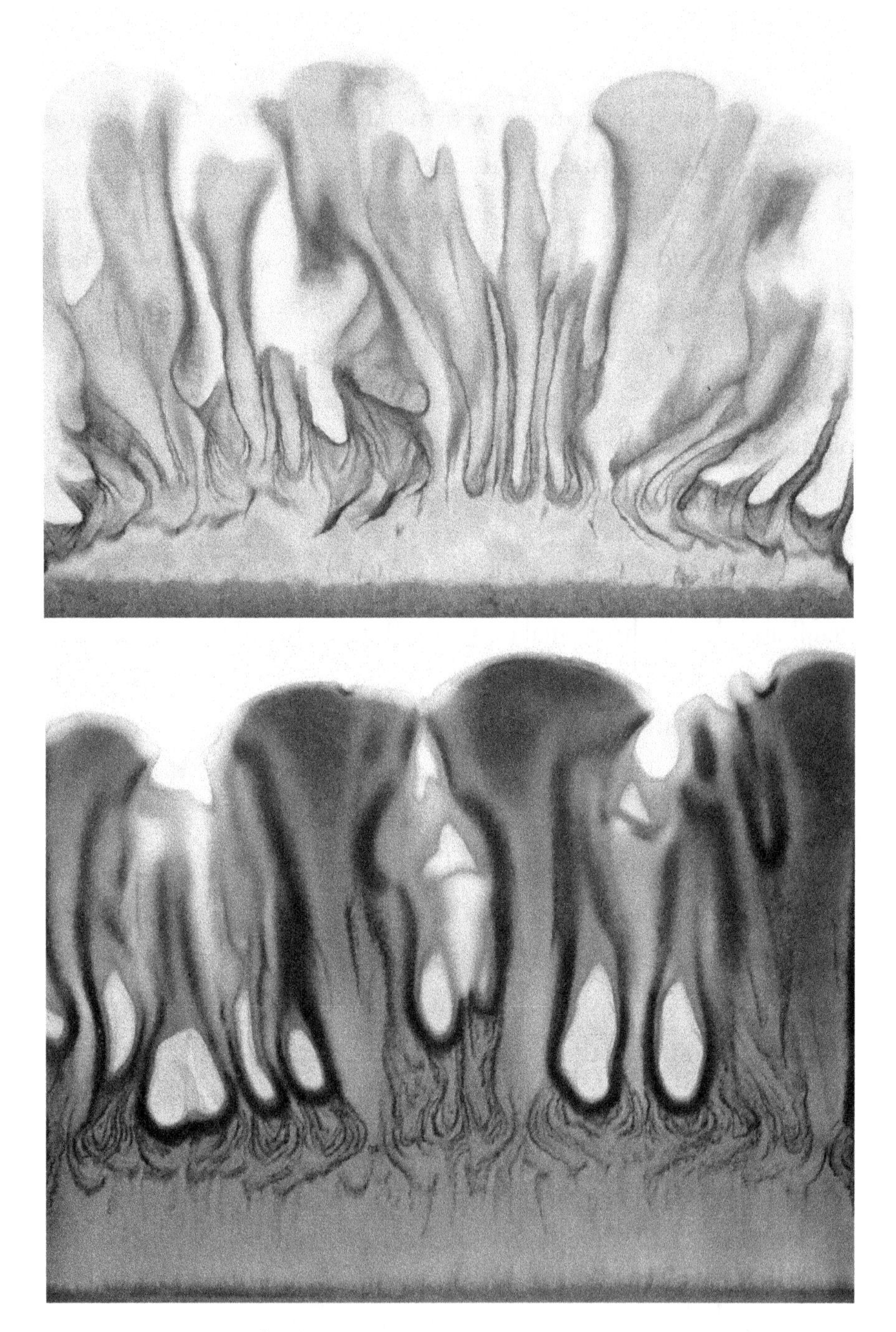

Figure 5.6a and 5.6b: Images created by capillary dynamolysis using mistletoe sap with gold chloride (top) and silver nitrate (bottom).

movement and uncomfortable in Germany they moved to England in 1934. Lili Kolisko established a small laboratory at the Old Mill House at Bray-on-Thames, the headquarters of the AAF, where she continued her research. In November 1939, Eugen Kolisko returning to England from a trip to Germany, collapsed at Paddington Station and died of a heart attack. Now on her own, Lili pursued her research with a single-minded focus. After the war she moved to Gloucestershire and set up a small laboratory in Brookthorpe at Wynstones School before moving to the nearby village of Edge. During this time she translated much of her husband's work into English. She published a substantial biography about him[7] and then a book they had been preparing together called *Agriculture of Tomorrow*. This major work brings together most of the research they carried out in relation to agriculture, although sadly it has not received the recognition it deserves.

In the second chapter of *Agriculture of Tomorrow*, Lili made a statement about the future task of agriculture. 'The agricultural problem is a world problem,' she declared, before going on to ask what use would it be if only one country were to produce healthy crops and its neighbours could not do the same? She emphasised the importance of a global approach and then showed how important it is for the three fields of agriculture, medicine and education to work together. They should not only stand side by side and support each another but should 'melt into one another becoming a living entity – world agriculture.' The food grown should be of such vitality that it can build up the human body and provide the strength to develop mental capacities. Medicine should be based on a true understanding of the human being as a threefold organism, as expressed in the nerve-sense system focused in the head, the rhythmic system of breathing and circulation centred in the breast, and the metabolic-limb system with its seat in the lower body. Furthermore, it should acknowledge the human being as having a soul and spirit as well as a body, for as she writes:

> A medicine which only treats the body ... is as harmful as an agricultural science which only looks at the chemical constituents of the soil and forgets life.[8]

Lili then summarised the tasks of the three sections of this world agriculture. The task of the Agriculture Section would consist of practical farming advice, agricultural research and education; that of the Agricultural-Medical Section, guidance on nourishment and diet, public health, medicinal herb cultivation and animal health; finally, the Agricultural-Educational Section would have the task of teaching children about food, nourishment and plant cultivation, but most importantly the task of educating the public.

To bring all this about, Lili believed, required resources and many collaborators but above all an enthusiasm and a love for humanity. This great clarion call for a world agriculture that recognises the spiritual in both the human being and the world, and incorporates medicine and education, looks towards a future that cares for and brings healing to the earth and all its people.

This vision is one that reinforces the sense that biodynamic agriculture is a social movement as much as an agricultural approach. It is also one that expresses deep concern for the future of humanity. Lili Kolisko emphasised this in her conclusion to *Agriculture of Tomorrow* in a comment that could not be more pertinent to today's conditions:

> What can be done to wake up mankind? This science is of a purely destructive character. It is pure intellect without a heart beating for mankind. It is a priceless jewel fallen into the dust. Scientists have to make the first step to redeem it again, to create a new science of life which places Man in the centre, which looks at everything from the standpoint of Man and not from the standpoint of science. Man has to be understood as a spiritual being.[9]

Lili Kolisko's contribution to biodynamic agriculture brought together in this work remains a rich source of valuable information for all biodynamic researchers, even though many of her experiments have not been, and in some cases cannot be, repeated.

6. The Organic Midwife

During the 1930s there was a growing concern over the direction mainstream agriculture was taking towards farm rationalisation (the cutting of costs and labour), mechanisation and fertiliser use. It was a period in which new values were being questioned and the loss of old ways was being mourned. As was often still the case at that time, inspiration for a different approach came from the aristocracy who owned large tracts of land. It was not a coincidence therefore that figures such as Gerard Wallop, the sixth Earl of Portsmouth (who styled himself Viscount Lymington), Walter James, known as Lord Northbourne, and, of course, Lady Eve Balfour, played such a prominent role in developing what would ultimately grow into the organic movement. A key role in this development was also played by the biodynamic movement and the personalities connected to it.

Lord Northbourne had long been interested in biodynamic farming. With the blessing of the Biodynamic Association, he arranged a nine-day Summer Conference on Biodynamic Farming on his estate in Betteshanger, Kent, in June 1939 – exactly fifteen years after the Whitsun conference in Koberwitz.[1] Some forty or so people with a wide range of interests were invited to attend the event. Leading biodynamic speakers included Ehrenfried Pfeiffer, Hans Heinze, who was then running the Loverendale biodynamic farm in Holland, Otto Eckstein, who conducted research into formative forces at the Goetheanum with Pfeiffer, and Dr Scott Williamson, the founder of the Pioneer Health Centre in Peckham, London. Besides lectures, farm visits were also arranged to the nearest biodynamic holding, an intensive mixed farm, a large-scale mechanised farm, and the University of London's teaching farm at Wye College (where Lord Northbourne was governor). There were also practical demonstrations of the biodynamic treatment of fruit trees. Unfortunately, due to the split in the biodynamic movement, members of the AAF and even the Experimental Circle did not hear about this event until afterwards.[2]

Lord Northbourne strongly believed that biodynamic agriculture offered a real alternative to the chemical-based agriculture that was so clearly harmful. In 1936, a few years before the summer conference, the idea was born of setting up a field experiment to compare the methods used by Ehrenfried Pfeiffer and Sir Albert Howard and those using artificial fertilisers. The plan was to use 100 acres (40 hectares) of farmland, take five fields growing five crops and use three management systems across the plots. The project was a joint venture

Figure 6.1: Walter James, Lord Northbourne.[3]

involving Ehrenfried Pfeiffer, Sir Albert Howard, Lord Northbourne, the Earl of Portsmouth and George Stapledon. It was known as the Farleigh Experiment and was due to run for seven years from 1936 to 1942. In the end it only ran for two years due to lack of money and the outbreak of the Second World War. It was, however, the precursor to the Haughley Experiment, the first proper comparison of organic farming and conventional farming developed by Lady Eve Balfour.

Figure 6.2: Lady Eve Balfour.

Lord Northbourne was also a writer and in 1940 he published *Look to the Land*, a book that was highly recommended by the Biodynamic Association. It was essentially a new landsman's philosophy. In it he described much of what he had learnt from biodynamic agriculture. He wrote about sustainability and the importance of the law of return. He emphasised the value of developing a farm organism, creating diversity and understanding how a healthy soil leads to healthy nutrition. His statement, 'The farm itself must have a biological completeness; it must be a living entity, it must be a unit which has within itself a balanced organic life', is almost a paraphrasing of Steiner's words in the Agriculture Course. He also wrote that:

> Every artist must first be a craftsman. Farming is the craft side of the art of living; that which we seem to have lost. With it has gone what Sir George Stapledon calls 'the spirit of place'; that which has made England the lovable land she is.[4]

The book provided the philosophy upon which organic agriculture would ultimately develop, and it was Lord Northbourne who first coined the term 'organic farming'.

A certain parallel can be discerned here between Lord Northbourne in England and Count Keyserlingk in Prussia. Both were of the aristocracy and both loved the land. Count Keyserlingk traced his ancestry back through the folk soul of his country and so did Lord Northbourne. Thus, while Keyserlingk provided the context within which biodynamics could be born, Northbourne served as a midwife to organic agriculture. However, while Keyserlingk put himself at the service of a new spiritual understanding of nature that could build on and consciously renew the cultural traditions of the past, Northbourne sought to tap into and re-awaken an existing stream of spiritual wisdom that was fast disappearing.

The parallels between these two personalities are many and tangible, and yet they also differ subtly and fundamentally. The difference becomes clearer in the later writings of Lord Northbourne, where he aligns himself ever more strongly to the circle of writers known as the Perennialists or Traditionalists. These were thinkers who were committed to a holistic view of life and held the conviction that all the world's philosophies and religions have the same origin.

It is an idea inspired by Tibetan Buddhism and Indigenous American traditions and it was Aldous Huxley's *The Perennial Philosophy*, published immediately after the Second World War, that brought this renewed understanding of our connection with the earth to a wider public. Biodynamic philosophy shares the same source and observes the same principles. It differs subtly, however, in that whereas the aim of the Perennialist is to regain harmony and adjust to the existing beauty and diversity of nature, recognising it as the ultimate and complete creation of a divine world that cannot be improved upon, the biodynamic approach, while still building on the cultural inheritance of humanity, aims to refashion nature ever more strongly in the image of the human being and to transform wildernesses into cultivated landscapes. In doing so the human being is encouraged gradually to take on the role and continue the work of the creator. The difference between these two approaches is therefore both subtle and fundamental, and it is perhaps this more than anything else that prevented Lord Northbourne from fully identifying with the biodynamic movement.

However, as John Paull points out:

> A direct unbroken lineage can be drawn from Koberwitz, Poland, via Dornach, Switzerland, to Kent, UK; and from Rudolf Steiner, via Ehrenfried Pfeiffer, to Lord Northbourne.[5]

Northbourne's *Look to the Land* drew on much that he had learned from Ehrenfried Pfeiffer, but also remained wedded to a traditionalist approach that provided a philosophical underpinning of the new organic movement.

The precursors of sustainable agriculture are to be found in the rural farming communities across the world. In the early twentieth century, when the British Empire was still at its height, there were a number of great agriculturalists, many of whom were sent out across the world to 'teach the natives' modern efficient methods, but who, in attempting to do so, came to realise the treasures of wisdom that already lay hidden in the practices of indigenous farmers.[6]

This is what happened to Sir Albert Howard during British imperial control of India when he was sent to teach the local population about modern agricultural practices. He soon found, however, that he had much to learn from them. He was able to tap into thousands of years of farming tradition, of working with and re-using organic materials in harmony with the whole of nature, and he learned of the self-renewing power of compost and soil humus. This experience led him to develop the Indore composting system. This 'gift from India' confirmed the contention of Lord Northbourne that those 'who handle the soil must look upon it and treat it as a living whole, for it is a living thing.'[7] It facilitated the birth of the organic movement in the British Isles, enabling it to grow strong roots and flourish through an organisation dedicated to the soil: the Soil Association.

'The divinity within the flower is sufficient of itself'

Another person who was instrumental in the foundation of the organic movement was Maye Bruce, whose great passion was making compost. Making compost lies at the heart of all true gardening and farming activity, and this was especially true for the newly formed organic and biodynamic movements. Ehrenfried Pfeiffer had written a lot about the value of compost, as had Albert Howard and many other organic and biodynamic pioneers. Compost making was taken up enthusiastically by biodynamic gardeners and it was their engagement with the process and the use of special preparations that first inspired Maye Bruce.

Maye Bruce was born in Cork, in 1879, and spent the first years of her life in Dublin. She was the eldest of seven children. Her father, Samuel Bruce, came from a well-to-do northern Irish family that had accrued considerable wealth from Belfast's largest whiskey distillery of which he was a director. His family came from a long line of Scots Presbyterian ministers who arrived in Ulster during the early seventeenth century. The Bruce family traces its origins back to Robert the Bruce, King of Scotland. Maye's mother, Julia Colthurst, came from Cork. Her family had once owned Blarney Castle. In 1884, when Maye was five years old, the family moved to Norton Hall in Chipping Camden, Gloucestershire. This was partly to escape the growing sectarian tensions in Ireland, but also because of Julia's growing involvement with the London arts scene.

Little is known about Maye's childhood except that she was educated at home and taught by a German governess. She loved the countryside and had a great interest in nature. She collected flowers and butterflies and enjoyed walking and mountaineering. Hunting was also something she enjoyed. She became a very good photographer and loved painting and drawing. During the pre-war years she became politically engaged and sought to encourage a strong sense of patriotism. In 1913 she joined the National Service League. She assisted its campaign for compulsory National Service training and gave a speech bemoaning the 'want of discipline amongst young men'. She believed 'Those feckless lads should be put through training which would teach them discipline and self-respect.'[8] At another meeting of the league in Huddersfield on December 5 of that year she declared:

> If we are not prepared to fight we should go home and look in our children's eyes; then pray god to kill them. After which we should build a monument to our cowardice from their gravestones.[9]

Such sentiments sound strange to us today, but jingoism of this kind was rife in pre-war Imperial Britain.

Maye's parents moved to London in 1912, leaving her to take on the family home. When war broke out two years later she set up Norton Hall as one of the first VA hospitals in the country. It was staffed by volunteer nurses and remained open until 1919. She was its commandant and received an MBE for her work. She then became involved with the Girl Guides, serving as County Commissioner for Gloucester, a position she held for the next seventeen years.

Maye sold Norton Hall in 1921 and used the proceeds to buy her own property: Hill House, a Cotswold farmstead in Sapperton, near Cirencester. The house sat on a beautiful hilltop location and had a neglected garden that Maye's ambition was to make productive. Her first task was to improve the poor and stony soil. She used an old pile of farmyard manure to help increase the fertility of the soil, but that eventually ran out and she had to find another approach. She had instinctively ruled out using chemical fertilisers.

It was at this point that Maye came across the newly founded Anthroposophical Society and learned about its approach to the soil and the value of making compost. This in turn led her to the Experimental Circle and what was to become the future biodynamic movement. Maye started working with the biodynamic preparations and, after a short while, found that she could produce a highly active compost that totally transformed the soil in her garden. After a few years her garden became a shining example of a successfully working biodynamic garden. Maye became actively engaged in the work of the Experimental Circle and served on the council of the AAF between 1933 and 1937. So convinced was she regarding the value of compost and of the biodynamic preparations that she sought to encourage their use in whatever way possible.

In the early years of the biodynamic movement the Experimental Circle sought to protect the preparations in accordance with the rules that had been agreed during the Koberwitz course. As mentioned in Chapter 2, one of those rules stated that:

> Only those members [of the Anthroposophical Society] should be admitted who (a) are in a position to carry out practical experiments in their own farms and gardens and in such a way that it is impossible for non-members of the Experimental Circle to ascertain how the preparations are made.

Adherence to this rule by the Circle and the council of the AAF was a source of great frustration for Maye Bruce. It meant that she could not discuss the preparations and how they were made with the increasing numbers of people who approached her but were not yet part of the movement. Much as she tried she was not able to persuade her colleagues on the council to have a change of heart. They felt duty-bound to protect these preparations and prevent them falling into the wrong hands.

Maye then started thinking deeply about the preparations. She knew how effective they were and how they had transformed her garden, and yet she was at a loss to know how this experience could be more widely shared. Discussions on the council could not resolve the issue and so one day she decided to research an alternative approach. She thought long and hard about it and one morning she woke up with the inspired thought that 'The divinity within the flower is sufficient of itself'. She decided to work with the herbs proposed by Rudolf Steiner, but instead of preparing them as he suggested she took a different approach, one that would allow her to speak freely about them and how they were made. She extracted the juice of each preparation plant – yarrow, chamomile, nettle, dandelion and Valerian – and made an infusion of oak bark, and then, by a process of trial and error, sought to discover the most effective dilution. This she eventually found to be a dilution in water of 1:10,000. These solutions, together with a seventh, a 1:10,000 solution of honey, were then used to treat her compost heaps in much the same way as the biodynamic preparations. The result was a rapid fermentation that produced a sweet and stable compost. A conventional analysis of the compost showed the quality to be almost identical to classical biodynamic compost.

Gradually, Maye refined the method for making this compost starter. The herbs are dried and individually ground down to a powder using a pestle and mortar, then they are sieved. Honey and lactose, or milk sugar, are then also ground for one hour to ensure a thorough and even mixing. The ground herbs are then added, mixed together and stored in an airtight jar to prevent moisture absorption. Immediately prior to use they are diluted 1:10,000 and inserted into holes in the heap. Because of its quick-acting effect on the decaying plant material it became known as the QR (Quick Return) Compost Starter. This compost starter is still being used by organic gardeners across the UK.

Maye found assistants to help her test out the QR recipe, including a nun from the Benedictine Order at Stanbrook Abbey in Yorkshire.[10] The convent had a thriving herb garden where compost making had been accepted practice for many years. Reports of the benefits of the QR starter spread to other convents of the Order and it was also taken up in other countries. The nuns of the long-established

Benedictine nunnery in Fulda, Germany,[11] were known for their close attachment to the monastery's gardens, which their predecessors had looked after for hundreds of years. After the war they took in a large number of refugees and in order to feed them they had to cultivate more land. They were in regular contact with other monasteries of the Benedictine Order, including Stanbrook Abbey. In one of the letters received in 1948, there appeared the following:

> Each generation is the steward of the soil upon which they live; it must therefore leave the soil to its successors in better heart than they found it. God, the creator of plants, animals and human beings and the seeds for future growth bequeathed to them, had made the earth capable of renewing itself indefinitely using natural materials.

Then later in the document:

> Organic materials must be returned to the soil because chemical fertilisers will only impoverish the soil.[12]

From then on the nuns decided to adopt organic practices. From Stanbrook Abbey they learnt about the QR starter and began using it in their compost. They took up direct contact with Maye Bruce in 1948 and in 1953 began manufacturing it for themselves and selling it to other gardeners under the name of Humofix. They have continued to do this ever since.

Maye Bruce published two books on the subject, *From Vegetable Waste to Fertile Soil* in 1940 and *Common Sense Compost Making* in 1944, in which she described how she developed her compost starter and its method of production. The success of the QR starter inspired

Maye to link up with Lady Eve Balfour and others to found the Soil Association and launch the Haughley Experiment. Jocelyn Chase of Chase Organics subsequently took on the manufacture and marketing of the QR starter.

Because Maye parted company with the AAF in the late 1930s and set off on her own, some have seen this as a betrayal of the biodynamic impulse. From her point of view it was nothing of the kind and she continued to retain the highest regard for biodynamic methods. Maye learned what she could of biodynamics and then connected it with her own understanding of the world. She also parted on good terms with her biodynamic colleagues. Her ultimate aim was to carry what she understood to be the precious impulse of biodynamics out into the world and reach as many people as possible. She was very clear, however, that her compost starter was something different from the biodynamic preparations. The question remains as to whether the QR starter can replace the biodynamic preparations or whether their function is somewhat different. The fermentation of these medicinal herbs within the earth and inside animal organs that are chosen for their specific functionality and relationship to the cosmic and alchemical environment, has a different and possibly more fundamental effect on the soil and plants. According to Steiner these give plants an inner strength and sensitivity that can awaken them to their cosmic environment. The success of QR as a compost starter and activator, however, clearly indicates that the two approaches can complement one another.

Maye Bruce remained active on the Council of the Soil Association through to the late 1950s and is counted among those who sought to carry the impulse of biodynamics into the Soil Association. Her life's goal was to 'give life back to the soil and thus eventually abolish disease in plant, animal and man.'[13] She died in 1964 aged eighty-five.

7. Ehrenfried Pfeiffer

Few people have influenced the development of biodynamic agriculture in the English-speaking world so strongly as Ehrenfried Pfeiffer. He was born in Munich in 1899. His father was an active army officer who died when Ehrenfried was only five years old. His mother then moved with him first to Schwäbisch Hall and then Nuremberg where he went to school. Ehrenfried's mother was an avid reader of philosophy, attended many lectures given by Rudolf Steiner and held regular study groups in her house. Steiner was also a frequent visitor and on one occasion when Ehrenfried was suffering an acute infection, recommended an effective treatment that possibly saved his life.

Ehrenfried's mother, who had a somewhat nervous and emotional disposition, found it hard to cope with her only child. He therefore spent a lot of time with grandparents who lived nearby and only spent the weekends with his mother. Ehrenfried's grandfather was a chemist who had a great love for nature, and he would take Ehrenfried out into the countryside and share his knowledge of plants with him. There wasn't a plant in the vicinity that his grandfather could not tell a story about. Although retired by the time Ehrenfried came to stay with them, his grandfather had a small laboratory where he imparted the magic of chemistry to his grandson, showing the changes in colour and form of the various substances in the retort. It was above all through his grandfather that Ehrenfried's love of science and nature took root. Throughout his school days he would spend his free time out in nature, exploring the woods, wandering down country lanes and communing with nature beings. He knew first hand of the elemental powers that assisted growing plants, those that created thunder and lightning, and even those at work in his own body.

For Ehrenfried, school life in Nuremberg was a real trial. He hated every moment of it. He made no friends there and was frequently alone. He also found the lessons boring and could make no connection with his teachers. His only interest came through the books he read for himself, often at night by candlelight, which he later believed was the cause of his poor eyesight. By the time he was fourteen, Ehrenfried had experienced loneliness, lost faith in what it is to be human (apart from what he experienced in the loving environment of his family) and declared himself an atheist. It was only thanks to his family, especially the loving personality of his grandmother and the free space and guidance offered by Friedrich Rittelmeyer, the first priest of The Christian Community who guided him through confirmation, that he was able to recover his faith in humanity.

When he was about ten years old, Ehrenfried showed a talent for music and so his mother arranged for him to learn the violin and piano. Ehrenfried was one of those rare students with perfect pitch and an excellent memory for learning pieces by heart. These gifts enabled him to attend a music college alongside school. Ehrenfried loved music; it gave him a new way to express himself. Music lessons, however, with their emphasis on mechanical learning and a teacher who appeared to have little understanding for his needs, became rather trying and so at the age of seventeen Ehrenfried gave up what might have led to a brilliant musical career. His sensitive ear remained, however. This helped him not only to appreciate music but also the inner depth of human speech. He often described how he could recognise in the intonation of a person's voice something of the inner being that was speaking.

When he was fourteen years old his mother married Theodor Binder, an anthroposopher with a leading position in the engineering company Bosch in Stuttgart. Through him, Ehrenfried learned about technology, engineering and life in a factory. When the First World War broke out he registered as a conscientious objector and applied instead to work in the factory. This work provided him with his first independent income. When the war was over he continued at the factory while studying physics and electronics at the Stuttgart technical college. It was at this time that he became aware of Rudolf Steiner and anthroposophy. His mother had never spoken to him about it as a child, wishing instead to leave him free to discover it for himself later. He attended a lecture by Georg Unger, a close co-worker of Rudolf Steiner, about courageous thinking in which he referred to Steiner's seminal book *A Philosophy of Freedom*. Ehrenfried later described how at that moment he forgot where he was and came face to face with himself and his destiny. In that moment he knew that:

> I must seek the man to whom thinking opens the portal of life, who demands courage in thinking and leads to the recognition of freedom.[1]

He went straight home and told his mother about his experience and she immediately gave him a copy of *A Philosophy of Freedom* to read. He read it from cover to cover and when he had finished he asked for more. His mother gave him *Knowledge of the Higher Worlds* and this, too, he read from beginning to end. He now knew the path he was to follow.

When Ehrenfried was nineteen, his stepfather was asked by Steiner to take on the management of the Goetheanum in Dornach, which meant moving the family to Switzerland. At Christmas the following year, in 1919, Ehrenfried saw the Goetheanum for the first time. It turned out, however, that it was not only his stepfather who was to work at the Goetheanum. With his new engineering qualification Ehrenfried was given the task of developing the Goetheanum stage lighting for which Rudolf Steiner had very specific ideas that had never before been implemented. Steiner wanted the stage lighting to be so mobile and fluid that it could express in pure colour the changing etheric soul-scape of a verse or piece of music that was being performed in eurythmy. It was a very challenging task but one that Ehrenfried understood and worked hard to realise. Over the next five years he became a very close colleague and friend of Rudolf Steiner.

On more than one occasion, Ehrenfried describes moments of intimacy during which he met Steiner not only as the great teacher

Figure 7.1: Ehrenfried Pfeiffer.

most of his colleagues and the world at large saw him as, but as a human being in all his vulnerability. He was present when tragedy struck and the Goetheanum went up in flames. He wrote in a very moving way how, 'When the fire broke out ... and destroyed this Goetheanum with all its beauty and treasures, part of Rudolf Steiner's being and body was destroyed.' Ehrenfried felt a deep compassion for the man whose life's work had in that moment gone up in smoke: 'I experienced on that night the human being Rudolf Steiner,' and he vowed then 'never to desert the master, nor his work nor the man, especially the man.'[2] It was above all Ehrenfried's loyalty and support that convinced Steiner that he must continue his work despite this great loss.

Ehrenfried continued with his scientific training in Basel during this time. His focus had been on physics but now, at Steiner's suggestion, he took up chemistry and included a wider range of subjects in his studies as well, among them biology, botany, geology and even psychology. He was particularly urged to learn about practical experimentation. To a certain degree it was a time of re-connecting with the childhood experiences he had had with his grandfather and which laid the foundations for his future scientific research work.

During this time Ehrenfried asked Steiner if it was possible to find a source of energy that is not based on destruction but which works with the forces of life. If so, could it be demonstrated and could it form the basis for a new altruistic technology?[3] Steiner replied that such a force exists, but has not yet been discovered. It is the force that makes plants grow and lives for example in the seed as seed-power. Steiner said that it is possible to create machines that react upon and are driven by this force. In order to demonstrate its presence, Ehrenfried began investigating etheric formative forces and developed the sensitive crystallisation technique.

One of his earliest projects, the details of which remain unknown, was to investigate these forces and how they might be applied. When he showed Steiner the results, Steiner said they revealed not the working of etheric but of astral forces. He said it was an indication from the spiritual world that the time was not yet ripe for making use of etheric forces. Only when threefold social principles are being practised on at least some territories will social conditions be such that these powers will not be misused. Until that time any further experimentation would be unsuccessful and should not be carried out.[4]

Work on sensitive crystallisation began around 1925. In the years that followed a laboratory was set up at the Goetheanum in the Glasshouse (Glashaus), the home of what eventually became the Research Institute at the Goetheanum (Forschungsinstitut am Goetheanum). Ehrenfried was joined by two colleagues, Erika Sabarth and then Erika Riese. They produced the first successful crystallisation pictures by intuitively finding the best method. Only later did they check through in painstaking detail to prove that their method was correct. Sensitive crystallisation is a method whereby precise amounts of copper chloride and the organic extract to be tested are dissolved in water. Drops of this solution are then placed in a glass dish and set to crystallise in a temperature-controlled chamber free of vibration. As crystallisation takes place a structured form emerges. This form reflects the vital activity of the organic extract and is made visible by the metal salt. The structural forms and patterns can then be read by the skilled observer. Sensitive crystallisation is one of the so-called picture forming methods that include capillary dynamolysis (developed by Lili Kolisko), circular chromatography and the drop-picture method. Ehrenfried and his colleagues spent many years at this laboratory exploring plant juices, blood and other substances using this method. It also came to be used as a diagnostic tool by doctors.

Figure 7.2: The Glasshouse at the Goetheanum where Ehrenfried Pfeiffer had his laboratory.

The purpose of carrying out sensitive crystallisation was, most importantly for Ehrenfried, a means to use this picture-building technique to develop an imaginative perception of living processes, to perceive and understand etheric forces. This imaginative perception required the training of a 'knowing eye', the development of an inner power of visioning and a capacity for conscious and imaginative inner judgement (*anschauende Urteilskraft*). There was, however, a temptation to use this technique in order to prove the vitality and efficacy of medicines.

Barbara Saunders-Davies, writing an appreciation of Ehrenfried Pfeiffer in the eighteenth issue of *Star and Furrow* in 1963, described her impressions of his work with sensitive crystallisation during a visit to his laboratory at the Goetheanum in 1935:

> Every morning there was a laboratory meeting with Pfeiffer in the 'Glashaus'. He used to study the previous day's crystallisation plates with all his collaborators around

him and dictate his diagnoses and reports. Dr Steiner had suggested to him that a possible way of studying etheric formative forces might be by means of a sensitive process of crystallisation. The young Pfeiffer had found by experiment that a solution of copper chloride to which a few drops of diluted liquid organic substances were added crystallised on a flat plate in varying and significant patterns. In the case of human blood, forms appeared from which the general constitution and many specific diseases could be diagnosed and even the affected organs determined. It was amazing to watch him read these plates and give his report and diagnosis. Much was fairly clear if one had studied his crystallisation method, but one always felt in studying the forms that his thinking reached a higher level which revealed more to him than any of his colleagues could achieve.[5]

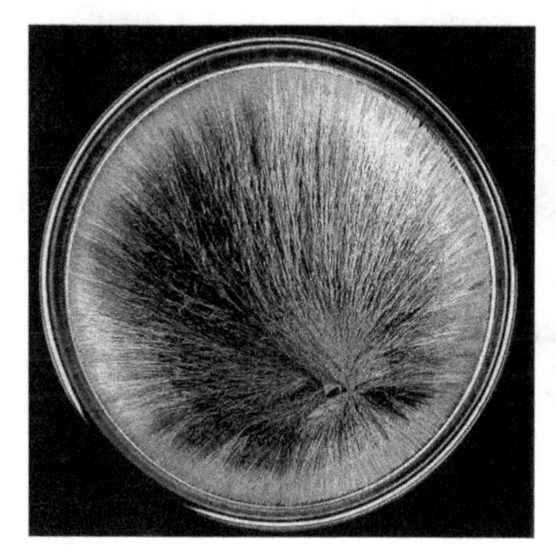

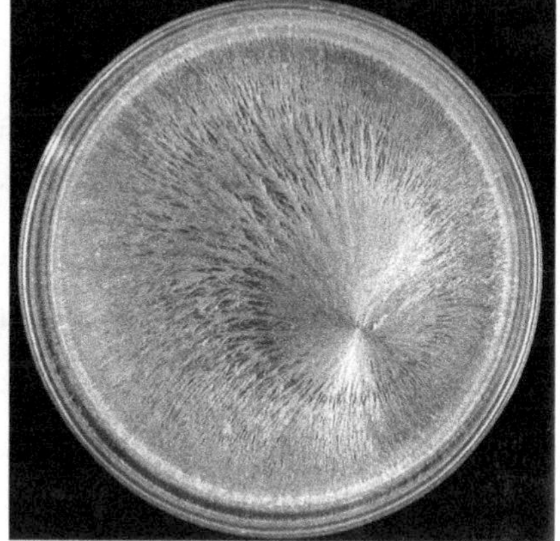

Figure 7.3a and 7.3b: Examples of pictures developed using the sensitive crystallisation with lime (left) and carrot (right).

In 1926, two years after the Koberwitz course, Ehrenfried joined a team of three directors to manage one of the first biodynamic farms in Holland. Loverendale in the west of the country was a farm of 500 acres (200 hectares) created from five smaller holdings. He was the farm manager for about ten years until, in 1935, he had to relinquish his post due to his commitments at the Goetheanum. This farm laid the foundation in a very practical sense for his life-long commitment to biodynamic agriculture.

He was a frequent visitor to Britain, often giving talks and workshops as well as serving as President of the Biodynamic Association from around 1939 until 1950, when it reunited with the AAF as the Biodynamic Agricultural Association (BDAA). He also attended the Betteshammer conference in July 1939. A short note in a printed report from this event also demonstrates his great sense of humour.

During a question and answer session someone asked: 'Could chamomile take the place of the customary spoonful of whiskey added to sugar for bees?'

Ehrenfried replied: 'I do not know. You can even get a male turkey to sit on eggs if you give him a strong drink. When he gets sober again he suddenly gets up and asks himself, "What have I been doing all this time!"'

He soon became a knowledgeable expert and shared his wisdom in many parts of the world, especially the USA where he travelled frequently during the 1930s giving courses and lectures. In 1940 Ehrenfried moved to America with his family at the invitation of a wealthy owner of an oil company. H.A.W. Myrin owned a 830-acre (335-hectare) farm in Pennsylvania that he wished to convert to biodynamic methods and develop as an agricultural training school. This was achieved with Ehrenfried's help, but after four years the working relationship fell apart.

Ehrenfried then decided to start his own dairy farm with his friend Peter Escher and so bought Meadowbank Farm, a 285-acre (115-hectare) holding in New York state. He put a huge amount of effort into developing the enterprise, but in the end it proved too much and his health suffered. He fell sick with TB and diabetes and had to leave the management of the farm to his wife; in the end, they had to give up the farm. In 1948, Ehrenfried was invited to Threefold Farm in Spring Valley to help establish a laboratory. He carried out research into many different fields including soil improvement, nutrition research, municipal composting as well as a study of plant properties (known as the doctrine of signatures) and the nature of substance. It was there, too, that he developed the well-known compost starter that is still being produced and applied today, particularly in North America. The Pfeiffer Starter, as it is alternatively known, is produced today by the Josephine Porter Institute.[6] It is a combined preparation that incorporates all the biodynamic compost preparations along with specially cultured micro-organisms (bacteria, fungi, enzymes) chosen to help the composting process. Like other compound preparations available in the biodynamic movement, it is particularly recommended as an easily applied compost starter for backyard gardeners. It was, however, originally developed for larger scale municipal composting.

Ehrenfried Pfeiffer continued his lecturing activity and advisory work throughout this period. Fourteen years after arriving in Spring Valley he suffered a heart attack and died on November 30, 1961.

8. Biodynamics in Post-War Britain

The tragedy of the split in the Anthroposophical Society and the trauma of the Second World War led to a renewed effort at reconciliation. Ehrenfried Pfeiffer along with others was instrumental in bringing about this reconciliation. Having worked with Rudolf Steiner at the Goetheanum he knew first-hand about the challenges and struggles that inevitably occur when a new spiritual movement is born. Indeed he had been caught up in these struggles and even spoke up in favour of the exclusions that occurred in 1935. As president of the Biodynamic Association, however, he always maintained a connection to the Anthroposophical Agricultural Foundation and was almost the only person with whom both the BDA and the AAF turned to when a speaker was sought.

The divisions in the Society during that period were of a very personal nature and the deep convictions held by the two parties meant any reconciliation was never going to be easy. From 1946 onwards, however, voices increasingly called for these two organisations to be brought together. Ehrenfried therefore penned a letter on behalf of the BDA inviting the council of the AAF to a meeting in London on September 25, 1950. This meeting of the two councils marked the beginning of a process that was ultimately to unite the two and bring about the birth of the Biodynamic Agricultural Association (BDAA).

The Biodynamic Association had seen its task as serving the needs of farmers and gardeners who knew nothing about anthroposophy but were interested in applying the biodynamic method. The Anthroposophical Agricultural Foundation on the other hand saw its task as being to deepen the anthroposophical understanding of agriculture. These two strands, one focused on practical work and the other on a deepened understanding of the spiritual foundations, reflect polarities that run through the entire anthroposophical movement to this day. Each one, however, is as important as the other.

Many other people worked alongside Ehrenfried Pfeiffer in his efforts to re-unite the movement. Foremost among them was David Clement who in 1951 succeeded in presiding over a reunited movement and becoming the chair of a new organisation called the Biodynamic Agricultural Association,* a position he held for nearly forty years.

David Clement was born in 1922 in Staines, Surrey, and received a private education first at Clifton College in Bristol and then at Kings School, Bruton. It was while at Bruton that he first came across anthroposophy. In 1930 he went to Oxford University to study history accompanied by his old school friend Adam Bittleston, who later went on to lead the Christian Community in the UK. Together they studied Steiner's work and met a number of other people who were active in the Anthroposophical Society, including Dr Ita Wegman, Fried Geuter and Michael Wilson. David's older brother, Eric, who was pursuing a career in the army, had also met and been impressed by Fried Geuter. He promised to join him at the newly founded centre for therapeutic education in Selly Oak, Birmingham, on his return from a trip overseas. Sadly, however, he died in India and did not return. Shocked and also disappointed

* Since 2012, the Biodynamic Agricultural Association has been known simply as the Biodynamic Association (BDA).

with life in Oxford, David abandoned his studies in 1932 and took up the post his brother had been promised.

The centre, founded by Fried Geuter in 1929, was the precursor of what later became Sunfield Children's Home in Clent. The move to Clent took place in 1933 and Sunfield soon became a leading centre for remedial education in Britain and, under the guidance of Michael Wilson, a focus for Goethean research throughout the middle part of the twentieth century.

David's work at Sunfield allowed him to deepen his studies of anthroposophy. He also became actively involved in the work of the Youth Section. It was during this time that he came across what was then known as anthroposophical farming and went to visit Sleights Farm in Yorkshire, where Maurice Wood was pioneering this new approach. David was very impressed and was keen to see it introduced on the small farm at Sunfield. Around this time, nearby Broome Farm came on the market and, using a timely legacy, David was able to purchase the land for the trust. The aim was to create a biodynamic farm that could supply Sunfield with milk and other produce. He initially asked Carl Mier to manage it for him and then, in 1937, it was taken on by Deryck Duffey. Broome farm was a 250-acre (100-hectare) holding about a mile away from Sunfield. Its soil in an area of terminal moraine and was described as being difficult to farm and very hungry. Despite this the farm was able to produce wheat and rye for making bread, as well as vegetables, eggs and dairy products (milk, cream and butter). The consistent application of all the biodynamic measures certainly helped build fertility. Between 1937 and 1940 Broome Farm and the Sunfield home farm formed what became known as the Sunfield Agricultural Centre. There was a farm shop and facilities to welcome visitors and those keen to learn more about biodynamic agriculture.

Figure 8.1: Broome Farm, Clent, in 1950.

David Clement continued working at Sunfield during this time and also joined the Experimental Circle. Then as now its purpose was to deepen the understanding of biodynamic principles. In those days it met quite frequently. In 1940 David married Fried Geuter's eldest daughter, Hilla, and a new phase of life began. That was also the year Deryck Duffey decided to leave the farm and move to Scotland. This new situation led Hilla and David to take on the farm themselves. In order to do so, however, they were required to purchase the farm from Sunfield and turn it into a family business. They achieved this with the help of family funds and for the next forty years they farmed the land and raised their family.

They were remarkably successful in producing high-quality produce. The dairy herd even won the second prize in the local farming competition. David continually sought to forge links with the local farming community and was elected three times as the chair of the local branch of the National Farmers' Union. He also became president of the West Midlands Shorthorn Society and for most of this time chaired the BDAA.

Figure 8.2: David Clement.

In 1986, David retired and the farm had to be sold, with the proceeds returning to the family trust. For many years Broome Farm had been seen in the UK as a good example of a working biodynamic farm, and so when it came on the market many people had a strong desire to save it for the movement. A huge fund-raising appeal was launched to buy the land, but despite almost succeeding, the appeal lost out on the final bid. With a great deal of sadness, members and supporters of the biodynamic movement had to acknowledge that the end of an era had come. David Clement continued to be involved with many initiatives close to his heart, including the Experimental Circle, Sunfield and Elmfield Rudolf Steiner school. He lived in a cottage in Clent to the ripe old age of ninety-seven. He died at home on May 20, 2004.

Meanwhile, in Scotland, a new biodynamic farming initiative was being started by Grange Kirkcaldy. When the First World War broke out he was nineteen and joined the army. He served throughout the war and continued with the army until 1932 when he retired and purchased an old manse with a large walled garden in Aberdeenshire. He renamed it the Lodge of Auchindoir. A year or two later he leased some further land and began farming it. Farming was new to him, but he took it up with enthusiasm. He had already discovered anthroposophy and the biodynamic approach and resolved to run his farm accordingly. His main focus was on dairy farming and the production of raw milk. This proved a success and he was able to expand the enterprise to a further 120 acres (48 hectares) of land, which he purchased some 20 miles (32 kilometres) away near Inverurie.

Figure 8.3: Grange Kirkcaldy.

In an Appreciation written for the *Star and Furrow* his wife, Mildred, wrote:

> Grange also made one of the 'do it yourself' experiments he and his children enjoyed. This was to make some silage themselves and see what the results were. A green crop of oats and peas was led in in a pony cart and chopped in a hand-driven chaff cutter with the help of two boys (or 'loons' as they were called in Scotland) to turn the cutter. The silage was stored in 4 ft diameter water pipes, sunk in the ground and consolidated by two small daughters. Black treacle was added, layer by layer, with a garden watering can. Grange was a great believer in black treacle for the health of human beings and of animals. A farmer's daughter once told him that on their farm the runt of a litter was always given black treacle. He liked to hear the old lore of the countryside and took delight in the country sayings and laconic wit of the North-East countryman.[1]

During the Second World War, Grange signed up as a reservist. This took him away from the farm, which proved hard to manage in his absence and in 1942 it had to be sold. After the war, Grange had planned to start farming again but instead sold Auchindoir as well. He continued living near Aberdeen and focused his energy on developing a centre for biodynamic group work and continued making the preparations till the end of his life. After his death in 1979 a legacy fund was set up to support the development of biodynamic work, the Grange Kirkcaldy Trust, which gave support to many new initiatives over the years. It is today managed by the Biodynamic Agricultural Association.

After the Second World War agriculture underwent rapid technological change. Tractors reigned supreme and all kinds of new machines were invented to ease the workload and make farming more profitable. So-called 'conventional' agriculture continued on its chemical and pesticide-based trajectory. This was encouraged by the introduction of a cheap food policy that inevitably saw a lowering of food quality. This led to a gradual public awakening and soon the problems of poor food quality, environmental damage and widespread pollution gained wider attention. Interest in the organic approach begin to grow more strongly. There was also increasing interest in the biodynamic approach.

In 1955, Carl Mier, who for so long had carried out the advisory work of the Biodynamic Association, decided that it was time to retire and so the newly constituted Biodynamic Agricultural Association had to find a successor. This was when George Corrin joined the management team. He became a much-appreciated consultant for all things biodynamic, and during the next twenty-five years travelled the country sharing his wisdom with farmers and growers. All the while he ran his own market garden, Tynewydd Ketch, in North Wales, which supplied produce for the local market. This was very important to George and essential to his work as an advisor as it ensured that he remained fully grounded and in touch with practical farming activity.

Between March 1961 and April 1985 George Corrin put together and mostly wrote the *Members Bulletin*, the predecessor of the current BDAA *News Sheet*. During these twenty-four years he produced fifty-two editions, eventually publishing three times a year in early spring, late summer and autumn. The articles in these bulletins were written in a direct and conversational style. There was a section called

'Leaning on the Gate with the Consultant', which contained valuable advice, humorous anecdotes and interesting information gleaned from news reports, research papers and information sent in by members. It sometimes included political comments such as the following, included here because it remains as true and pertinent today as it was in 1962:

> I'm not in the least concerned with any political party. Having recently had a by-election in Montgomeryshire, I remain unimpressed by all politicians and the statements of their parties. One thing they have in common: to provide cheap food for the people. If it can't be produced cheaply, it must be made to appear cheap! The price ticket on the food in the shop window is not the real price of the food; the real price would be very much higher. So the fictitious price is left on the product in the window and the difference is made up in the form of a subsidy. The taxpayer ought to work for his food and not expect it to be produced 'on the cheap' by another section of the community. Somewhere hasn't our sense of values got all mixed up? Food should not be cheap … It is the most important product in the whole of life and therefore the most valuable. Without food to keep us alive, what is the good of cars, TV sets, mink capes or even Leonardo cartoons? Either we realise food is the most valuable possession in the whole world and pay the price for it or we see that it is as necessary as the air we breathe and likewise make it free for all. For make no mistake about it: the aim to produce cheap food is one reason that quality has suffered. Over use of artificial fertilisers has produced

> watery crops lacking in flavour or the highest nutritive values. Cheap methods rule out free-range hens and lead to battery cages, latest research even suggesting that three birds should be put in one fifteen inch cage! High labour costs or the shortage of labour hastens on the use of weed killers. And so one could go on. A large number of questionable methods in agriculture and horticulture are not the result of farmers being wicked men but because they are trying to make a living.[2]

George was an avid reader in all fields and especially science. He also made an extensive study of anthroposophy. In doing so he often unearthed some pertinent quotes that he would later publish in the bulletin. For instance, the following statement by Rudolf Steiner in a lecture given in 1916 and published in a booklet entitled *Memory and Habit*:

> Strange as it may appear, in times to come, in order to understand the heavens, people will study the embryo (as it develops out of the cell) and its environment, up to the point of the existence of the human being as a complete and finished being. And the observations made will serve to reveal the mysteries of the great universe. The revelations of the heavens will be explanatory of processes which, on earth, take their course in animal, plant and human being – above all in embryonic life. The truth is that the heavens explain the earth and the earth the heavens. This still seems a paradox to the modern age but it is a principle of real knowledge for the future and one that must be amplified and developed in many directions.[3]

He also taught regularly on the biodynamic course at Emerson College, which was run by Herbert Koepf in the 1970s and 1980s. He offered a very practical introduction to horticulture and market gardening that drew largely on his own experience in Wales. This included, for instance, the six-year crop rotation that he used. He also emphasised the importance of having a full year devoted to fertility building crops. He described the rotation as follows:

Year 1: Fertility building crop
February/March sow vetch @ 1½cwt / acre
June when in full flower, crush and leave for 6–8 days and spray horn manure with barrel preparation and then work it into the soil once it is sufficiently decomposed.
Sow rye @ 1 cwt / acre and in late autumn work it into the soil, leaving ground open till spring.
Year 2: Early potatoes followed by late brassicas
Year 3: Late brassicas followed from May by salads and leeks
Year 4: Roots
Year 5: Autumn and winter greens
Year 6: Legumes

The lasting legacy of George Corrin is his booklet *Handbook on Composting and the Preparations.* This eminently practical guide was for many years given to every new member of the Biodynamic Agricultural Association and it remains an important and accessible introduction to the biodynamic approach. During the twenty-four years that George Corrin served as biodynamic consultant from 1961 to 1985, the Association experienced what could be described as a long period of stability even though membership remained static at around six hundred. David Clement was the Association's chair during that time and John Soper was its Treasurer until 1979. They became affectionately known as the three 'Broome heads', Broome Farm being considered the nerve centre of the BDAA during that period.

It was in the early 1980s that George Corrin took the first steps towards registering the Demeter symbol in the UK, which took over a decade to complete. After George retired as consultant in 1985 this work was taken on by Jimmy Anderson and his wife Pauline. Jimmy became the Association's Fieldsman and in this capacity began the process of developing what was initially called the Demeter Guild. This involved a lot of discussions with other organic certification bodies and the government. He retired from this work in 1997 with the scheme fully established and in 1999 was awarded an MBE for his services to organic agriculture.

Jimmy was born near Fife in Scotland in 1927. He was the son of a doctor and attended Edinburgh University with the intention of studying medicine. While he was there, however, he met his future wife, Pauline, and they decided to take on a farm together instead. They first farmed in Scotland and later moved to Dorset, where they started using organic methods and became interested in whole foods. Then as they searched for a suitable education for their children, they discovered Michael Hall School and Emerson College in Sussex. They moved to East Grinstead and took on Busses Farm – this time as a fully biodynamic operation – and also helped start a wholefood shop and cafe in Forest Row. Making the farm truly viable was a big challenge and, although they were very successful in cultural and social terms, it proved unsustainable and in the end had to be sold. Jimmy and Pauline then returned to Scotland and set up a shop and restaurant called Helios in the Grassmarket in Edinburgh. It was after this experience that Jimmy focused on advisory work both independently and for the Biodynamic Agricultural Association. Establishing the

Association's Demeter Certification scheme took up a lot of his time. In 1998 they withdrew from this work and set up Netherfield guest house in Dumfriesshire, which they ran for a number of years.

Figure 8.4: Jimmy Anderson with the Lifetime Achievement Award presented to him by the Biodynamic Agriculture Association in 2009.

Another key figure of this period was Herbert Koepf. He is best known in Britain as the author of several books on biodynamics but also as the director of the Agriculture Course at Emerson College between 1970 and 1990. He was born in 1914 in Herbrechtingen, a village to the east of Stuttgart in Baden-Würtemberg, Germany. From a young age he was closely connected with farming and studied agriculture at Hohenheim University. During the interwar years he worked on a number of biodynamic farms in the eastern part of the country. After fighting as a soldier on the eastern front during the war, Herbert returned to Hohenheim University and, in 1950, became a professor of soil science. As a scientist he was particularly keen to research the nature of the soil and the potential improvements to soil life that could be made through biodynamic practices.

Herbert's connection to Britain began in 1962 when he was invited by Francis Edmunds to join the faculty at Emerson College and start a course in biodynamic agriculture. For the next twenty years he ran a one-year study course with students coming from all over the world. His thorough scientific approach, extensive knowledge and dry sense of humour was much appreciated.

I was myself a student on his course at Emerson in 1978 and I could appreciate his very down-to-earth approach to both the soil and the spiritual hierarchies. I also remember his somewhat dry humour and how in our discussions about the value of manure the inevitable question came up about human faeces and whether they could be used on the farm. He would then look at us all and with a broad grin on his face sigh and say, 'Once we begin this topic, it will be hard to leave it.' There was indeed a long discussion before it was brought to an end.

From 1971 to 1988 he took on the leadership of the Agriculture Department of the Natural Science Section at the Goetheanum. In this position he was able to build links between the wider scientific world and the biodynamic researchers in various countries. He also wrote a number of books and translated many scientific papers into English.

In 1990, aged seventy-five, Herbert Koepf moved to the United States where he supported the research work at the Michael Fields Institute in Wisconsin. In 1994 he returned to his homeland as a resident of Nikolaus Cusanus Haus, a retirement home in Stuttgart, where he continued writing a book about the historical development of biodynamic agriculture in the twentieth century. Its focus is primarily on developments in the German-speaking world, but there is also a chapter on international developments.[4]

I first met Katherine Castelliz when she lived in the cottage on Tablehurst Farm in Sussex, surrounded by flowers and herbs growing randomly together. She knew her plants as she knew the animals on the farm. Dressed always in earthy browns and wearing a traditional-style headscarf rarely seen outside rural Austria, she moved among the cows and calves administering food, medicine or herbal supplements. Her knowing eye could see a problem before it appeared and she knew immediately what was needed. She was of true peasant stock and was a wise woman thoroughly imbued with anthroposophy. She had remarkable insights into many aspects of life and shared them willingly with those around her. Browsing through her book *Life of the Land*, in which she wrote down much of her experience, reveals many remarkable insights into nature and the spiritual impulses working through it as well as very practical guidance for farm and garden.

Katherine was born in Vienna in 1909, the second of three children. Her father was an architect and her mother an avid gardener who later dedicated herself to biodynamics and anthroposophy. Her early years were spent in the suburbs of Vienna. When the First World War broke out, her father was posted to the nearby mountains in order to safeguard the city's water supply. Despite the hardships and the struggle to survive, Katherine nevertheless described this period living high up in the Austrian Alps as an idyllic time. It was here among the flower meadows and by mountain streams that Katherine's love of nature was born. They kept goats and cultivated a vegetable garden. After the war she finished school and took up a training in horticulture, but after one year she transferred to agriculture – at that time a very unusual step for a young woman. The Agriculture Course had meanwhile taken place in Koberwitz, and it wasn't long before Katherine, supported by her mother's enthusiasm, found the approach to agriculture that was to become the focus for the rest of her life. After completing her training she worked on many farms in Austria before starting a new phase of her life at the Goetheanum in Dornach shortly before the outbreak of the Second World War.

Katherine worked in the Goetheanum research gardens where she participated in the plant-breeding projects carried out by Erika Riese and Ehrenfried Pfeiffer. She was particularly interested in the work being undertaken to develop new cereals from wild grasses.[5] She also became involved in the sensitive crystallisation work being carried out in the laboratory. This brought her into contact with Weleda and eventually led to her working for the company in Britain. Whilst there she was invited by Karl König to apply the skills she had gained with sensitive crystallisation to his medical work. She later spoke of how much she admired the dedication and absolute integrity with which König approached his patients. Medicine was not a field that Katherine felt particularly close to, and yet by observing and learning from König's remarkable diagnostic ability she was able to develop her own skills in veterinary work. Farming was her vocation, however, and when in 1956 König started the Camphill village community of Botton in North Yorkshire, Katherine immediately moved there and took on the small, marginal farm of Dalehead at the back end of Danby dale. There she introduced and cared for a small herd of Jersey cows and pursued the biodynamic approach with utter devotion.

Katherine was a very independent soul. She was someone who did not find community life easy and so she moved on and spent the next few years helping out on a number of biodynamic farms around the country. When Emerson College, which had started some six years earlier in the grounds of Michael Hall School in Sussex, acquired the nearby Pixton estate in 1968, she moved there and took on Tablehurst Farm with Jeremy Kent. She was much loved by the agriculture students at Emerson for her down-to-earth, practical approach combined with a deep spiritual understanding of the processes of nature.

In 1979 she moved to Wales, first to Camphill Coleg Elydir and then to Plas Dwbl. Here, at over seventy years of age, she finally took on a farm of her own along with her lifelong friends Barbara Saunders-Davies and Nim de Bruyn. Making a living on these poor, marginal soils in the Preseli hills was tough. They worked very hard and with the help of their faithful farmer, Keith Skipper, managed to create a remarkably harmonious biodynamic farm. It had a small herd of Welsh Black cattle, a couple of Jersey cows to provide milk, and a productive vegetable garden. There were even some field crops. The whole enterprise was held within the ownership framework of

Figure 8.5: Katherine Castelliz.

the Responsive Earth Trust, which was set up to promote biodynamic education and research.

With never a thought of retirement, Katherine took on the chair of the Experimental Circle in the last seven years of her life. Here she was in her element and was able to share with her colleagues in a wonderful way some of the deep insights she had gained into agriculture, nature, humanity and indeed the whole cosmos. The Experimental Circle was for her at the heart of all biodynamic work in this country and rightly so since it has existed continuously ever since biodynamic agriculture arrived here in 1928. Katherine died peacefully on April 20, 1997.

Another long-standing member of the Experimental Circle and many times Council member of the Biodynamic Agricultural Association was Anthony Kaye. He was born in 1925 into a family connected with a woollen mill in Yorkshire. He first encountered biodynamic agriculture on a training course with Deryck Duffy in Scotland, which inspired him to take up farming. Later, on a visit to Michael Hall School in Sussex, he met his future wife, Eve Brocklebank. One of Eve's relatives owned Kilmurray Farm in Ireland, which had been converted to biodynamics during the 1930s. Anthony and Eve took on this farm, and it was there that Anthony became interested in flour and bread making. He installed a flour mill designed by Maurice Wood to ensure high-quality flour. This eventually led to him moving to Forest Row and becoming the cook at Emerson College, where he pioneered vegetarian dishes that were hardly known at the time. After fourteen years at Emerson, Anthony and Eve returned to Ireland. Unfortunately, Kilmurray Farm had to be sold and so they purchased Inisglas in County Wexford. Anthony was also instrumental in founding the Biodynamic Agricultural Association of Ireland (BDAAI). In 1999 the couple retired to a nearby cottage but continued to be active in the biodynamic movement. When Anthony died in 2006, Eve moved to be with her daughter in Stroud where she continued as an active and interested member of the Experimental Circle until her death in 2007.

Figure 8.6: The Kilmurray Farm Estate in Ireland.

Figure 8.7: Eve and Anthony Kaye.

Alan Brockman also played a key role as a practising biodynamic farmer during the second half of the twentieth century. He was born in 1927 and grew up in Kent in a family that traced their local ancestry back to the fourteenth century. His father ran a village garage and kept a smallholding. When Alan was about twelve, just before the onset of war, his father took on the lease of a 120-acre (49-hectare) farm where they raised and milked cows, kept chickens and pigs, and grew wheat, oats and root crops. It was the beginning of Alan's love of farming. He carried fond memories of making hay by hand and harvesting the grain using a traditional binder, stooking the sheaves to dry, then loading them on to trailers with a pitchfork, carrying them to the stack and awaiting the arrival of the thresher. He always found it exciting to watch the heavy steam engine as it puffed its way slowly through the village to begin its work.

Machines and how they work had always fascinated Alan. This interest grew and on leaving school he decided to train in electrical engineering. It was during this four-year course that he met his longtime friend and colleague, Henry Goulden. The war was already over by the time they had completed their studies, but they still had to do national service, which they did together in Derbyshire.

Henry was the person who first introduced Alan to anthroposophy and it was while they were in Derbyshire that Alan had an opportunity to visit Michael House School in Ilkeston. He was inspired by what he saw there and from then on developed an ongoing interest in Steiner-Waldorf education. He was also able to join a regular study group on *Occult Science* with Alan Howard, one of the school's founding teachers. It was an experience that made a lasting impression on him and began his life-long study of anthroposophy. On another occasion Alan paid a visit to Sunfield Children's Home in Clent, near Stourbidge, where he met Carl and Gertrude Mier and for the first time learned about biodynamic agriculture.

Another trip took Alan and Henry down to Hawkwood College in Gloucestershire, which had just been established as a new centre for adult education and training. In its early years the college offered a foundation year in anthroposophical studies supported by arts and craft work. It was on that trip to Hawkwood in 1950 that Alan first met Ulrike Karutz, a young student from Germany who would later become his wife and the mother to their two sons Patrick and Leo.

After working for a couple of years for Siemens as an electrical engineer, Alan returned home to his parents in Kent. He had been experiencing problems with his health and this gave him time to recuperate. It also gave him time to think about his future and he soon came to the realisation that he should take up farming. After all, as he often declared: 'We cannot survive without a farmer to grow our food.'[6]

Alan started working with his father and they eventually decided that the time had come to buy a farm of their own and so, in 1953, they found and bought Perry Court Farm. For the next fifty years Alan was to farm this land, produce food, bring up his family and found a Steiner-Waldorf school in its midst. This biodynamic farm began life with 40 acres (16 hectares) of fruit trees, including Cox's Orange Pippin that were supplied to customers across the south of England. After about twenty years the trees had aged, the market was unpredictable, and Alan decided to change direction. The orchard was uprooted and the farm focused on producing beef, cereals and, later on, vegetables for the London markets.

Alongside his farming work Alan was engaged with the wider questions of food, agriculture and the world. From the time he started farming he was an enthusiastic participant in the work of the Experimental Circle. He continued as an active member of the Circle throughout his farming career and during the last twelve to fifteen years of his life he held the role of co-chair. He also did several stints as council member of the Biodynamic Agricultural Association and for a few years served as its chairperson. He became a regular and much sought after speaker on biodynamic farming and anthroposophy and was greatly respected throughout the movement. In 2008 he received a Lifetime's Achievement Award from the BDAA during a well-attended ceremony at Rudolf Steiner House in London.

Figure 8.8: Alan Brockman.

Alan was also active in the wider organic movement both locally and nationally. As a council member of the Soil Association he shared his many biodynamic insights and helped to develop the organisation's well known logo whose threefold form is designed to emphasise the interrelationship of soil, food and health.

Life on the farm was not always easy and they often had to struggle to make a living, but Alan knew that success does not come about overnight. Nor is it easy to bring new ideas to fruition. It took many years for the long-cherished goal of building a school on the farm to be realised. Alan had many ideas, some of which have been brought about while others are still waiting to be realised. Although frequently disappointed by the slow pace of progress, Alan was always positive. He looked patiently towards the future and carried within him the deep conviction that 'steadfast imagination will achieve all things'.[7]

9. Another Dimension

Biodynamic agriculture and its development in Britain owes a great deal to the Camphill movement, which in the latter part of the twentieth century created communities that cared for children and young people with additional support needs and provided sheltered living and a working environment for their adult lives. Farming, gardening and the care of the land always played a central role in these communities and every farm in Camphill sought to be biodynamic.

The story of Camphill begins with the biography of its founder, Karl König. He was born in Vienna in 1902. He was the only child of devout Jewish parents and his upbringing led him through the Jewish rites of passage, but as he grew up he gradually made a connection to Christianity through his own exploration and reading. He spent a lot of time on his own and was an avid reader. When he was twelve years old, the First World War broke out. Deprivation became more prevalent then and König found it hard to accept his family's affluence in the face of the poverty he saw on the streets. He preferred dressing in ragged clothes declaring, 'I am ashamed to go about the streets so well dressed when I see such misery in the people around me.'[1]

After the war he went on to study medicine. He encountered anthroposophy in his student years around 1921 but never met Rudolf Steiner – König later regretted turning down an opportunity to meet him at a lecture Steiner gave in Vienna. His close connection and engagement with the movement came about through his friendship with Ita Wegman, a medical doctor and close collaborator of Steiner who invited him to work for a while at her clinic in Arlesheim.

In the summer of 1928 he travelled to London and attended the World Conference for Spiritual Science arranged by Daniel Dunlop (see Chapter 3). Here he met many people in the medical and educational fields who would later play an important role in his life. As with Carl Mier this conference proved highly significant for his future destiny in Britain as well as for the biodynamic movement. After the conference, König returned to central Europe and became the school doctor at the newly founded children's home in Pilgramshain in Silesia – now part of Poland and not far from Koberwitz where Steiner's course on agriculture had been held. He remained there until 1936. During his time in Silesia König met his wife, Tilla, and started a family.

In March 1936, König moved to Vienna in order to start a medical practice. There he began to gather groups of young people around him who were eager to learn more about anthroposophy and explore how true communities might be formed. It was through these youth-group discussions and the lecture courses he gave that König first had the idea that would ultimately lead to the founding of the Camphill movement. The young people in the youth group enthusiastically engaged with these ideas and resolved to join him in putting them into practice.

When the Nazis annexed Austria in 1938, König was forced to flee persecution along with many of those who had been active in the Viennese youth groups. He went first to the Arlesheim clinic to see Ita Wegman. While he was there he received a letter from the British Consulate inviting him and a number of other doctors to come to Britain. He was encouraged by Ita Wegman to take up this offer and go to friends of hers in Aberdeen. She was also convinced that a curative centre should be established in Scotland to complement the one she had established at Sunfield in England. He set off, initially alone, for the comparative safety of Britain. His family would follow later.

In an essay written in 1959 and which later formed part of a booklet about the Camphill movement, König described his arrival in Britain and how he spent that first Christmas in 1938 in a tiny London hotel room 'alone, a drop in the vast human sea of a city, a stranger, a foreigner.'[2] Here in the light of a flickering candle he reflected that he was no longer in the land of Europe but in a country of the western world. Everything was new and strange. There was a dedication in a copy of the St James Bible given to him as a present that read 'to go forward with the confidence and resolution of a man in maintaining the truth of Christ and propagating it far and wide.' This pointed towards, and gave him strength for, the task lying ahead.

König settled with his family and young colleagues in a manse on the Williamson estate near Insch in Aberdeenshire, which belonged to the friends of Ita Wegman. Their farm was one of a number in the area that had discovered and begun to apply Rudolf Steiner's agricultural ideas. It was in a house on this estate that the first group of children with learning disabilities arrived and Camphill came into being. When the Second World War broke out in 1939, the men who were of German and Austrian nationality were interned as 'enemy aliens' on the Isle of Man. This left the women to pioneer this new community initiative on their own. With utter dedication, they established a home life for the children that allowed them to participate in the rhythms of nature and the seasons while sourcing the best food they could from the land in their care – which, initially, was just a garden.

Figure 9.1: Karl König, the founder of Camphill.

Meanwhile the men were living on the Isle of Man along with other so-called 'enemy aliens' from around the country, many of whom were engaged with anthroposophy. This period of internment provided an opportunity for inner contemplation, future planning and getting to know each other.

It was during this time that König made a deep study of Rudolf Steiner's Soul Calendar, a series of weekly verses that follow the soul's journey through the year and the interrelationship between the inner life of the human being and the outer world of nature. It is a calendar that can be seen as an accompaniment to the Agriculture Course, a means for entering more deeply into the cycle of the year. In the Agriculture Course Steiner speaks about the year's cycle and how, for instance, the mid-winter period between January 15 and February 15 is the time when in the northern hemisphere the earth is most within itself and awake, when the power of crystallisation is at its strongest. According to Steiner, this is the time when the farmer or gardener can fruitfully practice deep inner concentration and focus meditatively on the farm individuality to bring about a healthy balance, returning potentially harmful influences to their rightful domain. Something of this quality is expressed in the verses written for this period.

42: January 15–25
In this time of winter dark
There manifests as strong intent
The soul's own inner power,
To navigate benighted realms
And feel through warmth of heart,
The coming revelation of the senses.

43: January 26–February 1
In the cold and wintry depths
True spirit warmth is found;
A presence in the mirage of the world
Is forged by forces of the heart
And in defiance of the icy world
The human soul-fire burns within.

44: February 2–8
Astir with new impressions
The clarity of soul transforms
Through spirit birth fulfilled,
Universal, untamed, sprouting life
With my creative thinking will.

45: February 9–15
The power of thought takes root
In union with the spirit birth;
It lifts the dim sensations
To light-filled clarity.
If richness of the soul
Would meld with universal life,
Then outer revelation
Must receive the light of thought.[3]

It was while interned on the Isle of Man that König and his colleagues began to consider more deeply how a community could be developed. It was guided by a deep study of Steiner's ideas for a threefold organisation of society, but also the experiences of König's own Jewish upbringing and discovery of Christianity. A powerful dream of meeting Nikolaus Ludwig von Zinzendorf,[4] the founder of the Herrnhut community, led him to introduce the Bible Evening, a weekly community event that continues to play an important role in Camphill. He later described Zinzendorf as one of the three stars of Camphill. The other two were Amos Comenius,[5] who sought to establish a universal college of learning, and Robert Owen,[6] who set up the well-known cooperative at New Lanark in Scotland. The impulses brought by these three figures formed the core of what was to become the Camphill community. They also reflected in a unique way the threefold social impulse that Steiner had sought to introduce (see Chapter 1):

- Cultural sphere (learning and free initiative): Comenius's Universal college – Camphill College meeting.
- Rights sphere (equality and humanity): Zinzendorf's devotional brotherhood – Camphill Bible evening.

- Economic sphere (service and mutual support):
  Owen's New Lanark initiative – Camphill non-wage ethos.

The College meeting belongs to the sphere of learning. Each person draws on their own unique skills and qualities to address a particular challenge, thereby providing a broad range of perspectives and understanding for what is needed. The Bible evening is a social event during which people are invited to share their deeper thoughts on life and the world. Camphill's non-wage ethos in which no one receives a wage draws on Steiner's fundamental social law:

> In a community of human beings working together, the well-being of the community will be the greater the less the individual claims for themselves the proceeds of the work they have done themselves; that is, the more of these proceeds they make over to their fellow workers, and the more their own requirements are satisfied not out of their own work done, but out of work done by the others.[7]

These three elements are known as the three pillars upon which Camphill is built.

Camphill's unique community structure brought young people to Scotland from across Europe to work with and care for children and, later, adults with learning disabilities. Central to this was the focus on the land and the cycle of the seasonal festivals. The garden at Camphill House on Deeside was the first piece of land worked by Camphill. It was here that children took their first steps in cultivating, weeding and growing the produce that was so much needed to supplement the wartime rations. The Camphill gardens continue to play a vital role in supplying food and outdoor activities as well as providing a context for the important seasonal festivals that are so much part of community life.

In 1956 a new step was taken with the founding of the first village community for adults with learning disabilities. Botton Hall in the North Yorkshire Moors is a large estate at the top end of a dale. Taking on this desolate, marginal and originally treeless land was a challenge for the early pioneers, but over time its farms were able to supply the community with a steady supply of milk and vegetables as well as meat, while providing meaningful work opportunities for its resident villagers. The farming carried out in Botton was surprisingly productive and led to the establishment of food-processing operations and a cheese-producing dairy.

Figure 9.2: The garden at Botton Farm during the 1960s.

There were initially four separate farms, each one with a slightly different focus. Later, two more farms were added. For many years the Botton farms stood out as examples of biodynamic farming, and many different farmers worked and cut their teeth there. Other villages with thriving farms were subsequently established in England, Scotland, Ireland, Germany, South Africa and the United States.

The importance of the land for Camphill manifests itself in many ways. Annual land conferences that welcomed people from the wider biodynamic movement became occasions for sharing innovative ideas and deep insights. Already in 1947 at such a conference König explored the deeper aspects of biodynamic farming, such as the nature of the animal organs used for making the biodynamic preparations. What he brought then was utterly unique and built on the work he had done early on in his career on embryology while working with Ita Wegman in Arlesheim. He also spoke about the origin of domestic animals, the role of insects and the world of elemental beings.[8]

During another agricultural conference in February 1961, the idea of a 'land payment contribution' system was introduced that would meet the costs of caring for the land, but crucially the farmers would not rely on income from sales of produce. Instead they would receive a contribution in recognition of the care taken for the land and the food produced. The wider community would receive produce as a gift and those working the land would be supported by a gift from the community. This concept built on the idea at the heart of the Camphill impulse, namely that work and personal income are to be kept strictly separate. Work is performed as a loving service and in return the community meets the needs of the land workers. It was in effect the first conscious step towards what would later become the community supported agriculture movement (see Chapter 10).

The Camphill movement expanded first to England and then to Northern Ireland in the 1950s. During the latter half of the twentieth century new communities sprang up all over the world and they almost always included a biodynamic land initiative. Today there are Camphill communities in Austria, the Czech Republic, Estonia, Finland, France, Germany, Hungary, the Republic of Ireland, Latvia, Netherlands, Poland, Russia, Sweden, Switzerland, South Africa, Botswana, the United States, Canada and Vietnam.

10. Community Supported Agriculture

My own first contact with Community Supported Agriculture (CSA) was through Trauger Groh, author of *Farms of Tomorrow* and an early pioneer of the scheme. I worked at the time as an apprentice at Buschberghof Fuhlenhagen, the biodynamic farm in northern Germany that he co-founded during the early 1970s. The farm began as a partnership between three farming families. One of them had inherited the property and others brought capital to it. It was all placed in a Trust whose objectives were to research biodynamic agriculture and its social context. Responsibility for the farm was shared between the three partners: one focused on the dairy herd and creamery, another on cereals and bread, and the third on vegetables. Overall responsibility was carried collectively. Customers from Hamburg came out to the farm to collect raw milk, quark, butter, bread and vegetables. This was the beginning of what became, and continues to be, a very successful community-supported farm.

Some years later, Trauger emigrated to America and helped to found the Temple-Wilton community farm in New England. The success of this venture led to new CSA initiatives springing up across the country. Trauger set down the philosophy and principles for establishing community-supported farms, or CSAs as they have become known, and gave practical guidance in his books *Farms of Tomorrow* and *Farms of Tomorrow Re-visited.*

Figure 10.1: Trauger and Alice Groh.

How community-supported farms have been created varies considerably as does the way they budget their farm costs. In North America the following procedure is common. First, a clear and detailed farm budget is put together. Once all the budgeting has been done, a meeting is called for all the farm members. This usually takes place over a whole day and is like a little festival. The farm year is reviewed and plans for the coming season are shared. Then the farm group gives a detailed financial overview and presents the budget. Every detail can be questioned and looked at. At the end of the process the members agree the budget. Having agreed it, they have to find a way of meeting all the agreed costs through their collective contributions. A process then begins that is best described as 'auctioning the budget'. On the basis of a rough guideline (the amount of the budget divided by the number of members) each person pledges to pay a certain sum within the framework of what is individually affordable. There are no fixed pledges. This makes it possible for those on low incomes to pay less, while those who can pay more are allowed to do so. A fixed-share arrangement would prevent the better-off from contributing more.

If on the first round the total pledges do not cover the budget, a further round of pledges is made until the total budget is met. Some budget expenditure may also be reduced. This system seems to work well, especially since the commitment is for one year only. No one can commit themselves for longer than a year at a time. Payments are then usually made by monthly standing order.

CSA arrived somewhat later in the UK with a broader concept that included a whole range of local food initiatives ranging from farmers' markets and box schemes, through intentional communities like Camphill, to fully developed community-supported farms. In this context CSA is defined as a partnership between farmers and consumers, where the responsibilities and rewards of farming are shared.

One of the first CSAs to be established in the UK was Stroud Community Agriculture,[1] a biodynamic farm close to Stroud in Gloucestershire. It was founded in 2002. From a modest beginning in a walled garden it has gradually grown with a membership of around 280 receiving weekly shares of vegetables and farming around 50 acres (20 hectares) on two separate sites. It is structured as a fully constituted cooperative (the Industrial Provident Society) and members are encouraged to participate in the life of 'their' farm. Members' contributions support the whole farm and in addition to weekly vegetable shares they can also purchase meat.

Figure 10.2: The Stroud Community Agriculture's biodynamic farm in Gloucestershire.

Community Supported Agriculture is a system that builds on the concept of the biodynamic farm as a self-contained organism, meaning that whatever is needed for agricultural production should, wherever possible, come from the farm itself. This is fundamental to developing a strong internal cycle on the farm (from soil to fodder to livestock and back to soil), it builds fertility and strengthens the farm individuality, making it more resistant to disease. The grass, the cows, their intestinal flora and the soil micro-organisms, all adapt themselves to one another until, over time, an individual identity emerges that is utterly unique and site-specific to that farm. It becomes in effect a living being with a biography. This leads directly to the ideas behind community-supported agriculture – a farm conceived of as a living being is the foundation for a true community. Just as a healthy farm evolves out of its own resources, so can its 'surplus' provide for a community. The more diverse a farm becomes the more will it be able to meet the requirements of its community. For example, fruit, herbs, eggs and even textiles can be produced as well as milk, vegetables and meat. Diversity benefits both the farm and the community.

On the face of it, this sounds like simple self-sufficiency. There is, however, a fundamental difference. The goal of self-sufficiency is to become independent of one's neighbours and maximise personal gain. The gesture of a community farm, by contrast, is one of service to humanity.

The model for CSAs developed by Trauger Groh was partly inspired by experiences gained from the Camphill movement (see Chapter 9), which in turn sought to build on the threefold social order conceived of by Rudolf Steiner. During the period of global meltdown immediately after the First World War, Steiner tried in every possible way to encourage those in positions of responsibility to develop a more enlightened approach to the social question. He proposed that in place of the failed concept of a unitary nation state in which human rights, individual initiative and economic interests continually collide and undermine one another, society should be re-ordered so that these three elements have space to fulfill their true roles without being compromised.

According to the threefold social order, society can be differentiated into three distinct spheres of activity: the cultural sphere, which includes education, medicine and the arts; the state, which upholds human rights; and the economic sphere, which encompasses the production, distribution and consumption of commodities and services. The guiding principle of each is, respectively, freedom, equality and brotherhood. These are the ideals of the French Revolution, but far from being utopian, they take on very practical qualities when applied to the three spheres of the threefold social order. In the wrong context, however, these three principles can also be harmful: for instance when the economy gives free reign to individual greed rather than providing a brotherly service to the community, or when equality and uniformity places individual expression in a straitjacket, or when the state becomes a brotherhood of interest groups instead of promoting justice and equality.

In a CSA farm, the principle approach to work could be described as an enhanced form of volunteering. Involving as many people in the farm as possible is always a core aim since members are encouraged to view the farm as their own. The payments made by members support the farm and in effect enable the farmers also to volunteer since they don't need to sell the produce but can devote their time freely to the farm. Such an approach is not dissimilar to that of Camphill. It also corresponds to aspects of the thinking behind the threefold social order. The activity of farming is an activity of the free spiritual life sphere. Only when products are bought or sold is there an aspect of the economic sphere, while the relationships formed and arrangements made between farmers and members and the wider community express the sphere of rights.

More and more CSA schemes are being created today with many different forms and structures. Most of them are either organic or biodynamic and the focus is usually on vegetable production, although there are farms that focus on livestock, fruit or flowers.

11. More Recent Initiatives

During the 1980s a group of mainly younger biodynamic farmers and gardeners came together to re-invigorate the biodynamic movement in Britain. There was a perception that the movement was stagnating: conferences and events were becoming somewhat stale and younger people were not attending. There was also an issue about the annual conference being held in early July, a time when farmers in particular couldn't get away from the farm. Meanwhile, a change in leader of the Agriculture Department (then still part of the Natural Science Section) at the Goetheanum, brought with it a reinvigorating mood. This inspired a group of young farmers and biodynamic students to embark on a new initiative. Over the course of several meetings they developed the idea of holding a winter conference to explore the fundamentals of biodynamic agriculture in a different way and make it more accessible to the upcoming generation. This group became known as the International Biodynamic Initiatives Group (IBIG). It also had the intention of making a connection with the Goetheanum.

With this in mind, Manfred Klett, a founding farmer of the Dottenfelder farm community near Frankfurt and later head of the Agriculture Section at the Goetheanum, was invited to give the keynote lectures during the first conference in January 1985. It was so successful that Manfred Klett came to Britain to attend subsequent conferences for the next twelve years. Each conference took place during the New Year period and the theme chosen built on that of the previous year. Many of the lectures and contributions were subsequently published as small booklets.

1986 – Tomorrow's Agriculture
1987 – Building Stones for Meeting the Challenges
1989 – Growing Together: Why Should we Bother?
1990 – Dying and Becoming: Man's Path to a New Communion with Nature
1991 – Agriculture as an Art: the Meaning of Man's Work on the Soil
1992 – The Biodynamic Spray Preparations
1993 – The Biodynamic Compost Preparations
1994 – The Biodynamic Preparations as Sense Organs
1995 – The Sheaths of the Preparations
1996 – The Biodynamic Farm: Individuality and Community
1997 – Biodynamics: Cultivating Morality
1998 – The Future of our Cultivated Plants

Themes chosen for the annual winter conference of the Agricultural Department, during which Manfred Klett gave many lectures and keynote speechs.

What developed during these conferences with the help of Manfred Klett and contributions from other biodynamic farmers brought about a deepening of the biodynamic work in Britain. They have had a lasting effect. Just as importantly, they renewed the connection to the work of the Section at the Goetheanum.

Figure 11.1: Manfred Klett giving a geology lecture to students from the Agriculture School in 2009.

When I first went to the Goetheanum conference in the 1980s and 1990s, the event was held entirely in German. Although it was an international conference, participants from other European countries, Britain, America and elsewhere, had of necessity to speak German. This reinforced British insularity and the sense among biodynamic practitioners in Britain that the Goetheanum was far away and had little relevance to daily practical life on these islands. In raising awareness for the importance of the School of Spiritual Science at the Goetheanum, and with the inspiration he offered, Manfred Klett fostered a greater consciousness of the worldwide movement, which led to more people from Britain being interested in attending the conference in Dornach. As the new century dawned, things changed rapidly and people from across the globe began to attend the conferences. It wasn't long before people who had English as a first or second language made up a significant portion of the participants. The biodynamic movement had become truly global, and today nearly fifty different countries are represented at the conference.

Biodynamic seeds

The production of biodynamic seeds had been quietly developing for many years. In the 1980s and 1990s the growing threat posed by new hybridisation and genetic modification procedures gave extra impetus to the growing of biodynamic seeds. This led to the launch of a biodynamic seed project in the UK during the first years of the new millennium. Its aim was to encourage farmers throughout the country to start growing and saving their own seeds and ultimately to participate in a biodynamic plant-breeding initiative. Initiatives of this kind were well established on the continent and a key step was taken in this country with the founding of Stormy Hall Seeds by Hans Steenbergen, a farmer of many years standing in Botton Village, North Yorkshire. He was supported by a network of farmers and gardeners around the country who grew specific vegetable varieties on their biodynamic holdings. A catalogue of highly vital biodynamic vegetable seeds was then gradually established, and its quality has been increasingly recognised beyond the biodynamic movement. The initiative grew and developed and, in 2014, it became the Biodynamic Seed Cooperative.

By now it was outgrowing its Camphill village home and a new site was needed. A new and courageous step was then taken with the support of the Biodynamic Agricultural Association and other partners to purchase some land in Lincolnshire. Its new home, formerly Gosberton Bank Nursery, comprises some 24 acres (10 hectares) of land with 2½ acres (1 hectare) of glasshouses. The Seed Cooperative moved there in 2016 and is now well established. It has since become a leading supplier of organic and biodynamic seeds in the UK.

A huge amount of work goes into maintaining the many different varieties of seeds to ensure that they remain true and consistent. All the seeds are open pollinated, which means that growers can save their own seed. It also means that the full, vital potential of the plants is retained. As well as maintaining well-tried and proven varieties, biodynamic plant breeders are developing varieties that are particularly suited to biodynamic growing conditions and that have improved nutritional qualities. Entirely new and exciting opportunities in plant breeding are also emerging but always with the integrity of the plant in mind. Biodynamic plant breeders in different parts of Europe are improving and creating varieties by paying particular attention to the soil conditions, climate and cosmic constellations as well as the choice of sowing and harvesting times. New approaches are also being researched that use music, eurythmy or other rhythmic approaches at key moments in the plant's life cycle to influence its development.

Apprentice training

The biodynamic apprenticeship training programme, which is now well established, has been evolving since the 1980s. Several farms offered young people practical training, and to support them a series of week-long block courses was developed and arranged through the Biodynamic Agricultural Association. These provided the spiritual and theoretical background needed by future farmers. This initiative was carried for many years by Pat Thompson who was also editing *Star and Furrow* and organising events and conferences. There were four block courses, which took place over two years. Their thematic content was subdivided under four broad headings with a block being devoted in turn to soil, plants, animals and astronomy. Each year some 20–30 apprentices have participated in the programme.

Figure 11.2: Making the barrel preparation with apprentices at Fern Verrow Farm, Herefordshire, 2006.

A further element was added during the 1990s when Oaklands Park community (part of the Camphill Village Trust) grew its apprentice programme to form a major part of its land-based activity. Young people were drawn from all over Europe to participate in community life and learn from the inspiring tutoring of Tyll van der Voort. Participating in the course became a journey of self-discovery as well as providing a practical training. For many, working intensively with the land led to the start of a successful agricultural or horticultural career, while for others it led to something quite different, such as an artistic or social therapeutic vocation. The training developed further, with additional short weekend support courses arranged in the region and taught by land workers. The artistic work included in these weekends and in the block courses served to deepen background understanding as well as acknowledge the 'culture' of agriculture. The training has since evolved further and has now gained accredited status through the Crossfields Institute.

12. A Challenging Start to the New Millennium

I was sitting in the office of the Biodynamic Agricultural Association in September 2001, when the door opened and a friend rushed in with the news that aeroplanes had hit the twin towers in New York causing them to collapse into rubble and killing nearly three thousand people. This attack resulted in stringent travel restrictions and twenty years of relentless, ineffective war against terror. Just a few months earlier, Britain had been rocked by the foot-and-mouth epidemic, which had led to scenes of slaughtered cattle being burned in their thousands on vast funeral pyres.

Foot and mouth disease (FMD) has been around for a long time and major outbreaks have occurred in regular 30–40 year cycles: in 1967–68, in 1923–24 and in 1882–83. Abigail Woods, a student of veterinary medicine at Cambridge University, wrote a comprehensive analysis of the FMD phenomenon. In *A Manufactured Plague: A History of Foot and Mouth Disease in Britain*, she describes how the response to FMD evolved during the course of the nineteenth and twentieth centuries. When it was first observed in animals 150 years ago it was treated as a common, frequently recurring ailment that had simply to be lived with. In other parts of the world, too, FMD was seen as a mild affliction that is easily treated with natural medicines. It was commonplace for farmers to speed up the development of antibodies in their herds by encouraging cross infection as quickly as possible. Healthy animals given the cud of a sick animal would then become resistant without getting sick. This was a natural way of advancing herd immunity. There were also many natural and homeopathic treatments available that had proved effective in some places. FMD is therefore a disease against which most animals will survive and develop antibodies. High-yielding dairy cows, however, rarely regain their previous productivity and their rate of growth is often reduced. With the advent of more intensive farming methods and the breeding of highly productive cattle for the export market, this could no longer be tolerated and the attitude towards the disease changed. The state intervened and FMD became a notifiable disease that had to be eradicated. This policy continued right through to 2001 and beyond.

The FMD epidemic in 2001 was disastrous for the farming community, including many well-established biodynamic farms. The BDAA, along with all the other farming organisations, was deeply concerned about the consequences of the FMD epidemic and participated in many of the debates raging at the time as to whether cattle should be vaccinated or not. While many European countries chose to vaccinate cattle, the UK maintained a policy of eradication by slaughter. Both policy approaches, however, are based solely on materialistic considerations. From a biodynamic point of view it is clear that having healthy animals is dependent on developing a whole farm approach. Healthy soil, diversity of plant life, mixed cropping, home-produced feed and closed herds all contribute towards animal health and optimum levels of productivity. These, along with the healing effects of the biodynamic preparations and a more spiritual understanding of life, go a long way towards achieving resilience against disease.

A biodynamic treatment for foot and mouth disease was suggested by Rudolf Steiner and developed by Lili Kolisko in the 1920s, known as the coffee preparation. It built on thorough research into the symptoms of FMD, how the disease progresses and which parts of the organism need strengthening. They found that the disease process arises as the result of a disconnect between the nervous system and the rest of the organism caused by a disturbance of the rhythmic system:

> To strengthen the rhythmic system, to re-establish the connection when disturbed, to stimulate the circulation in the nerve-sense system and especially towards the brain, to restore the digestive activity to its normal condition – in short to repair the disturbed rhythm and re-unite the upper and lower systems – such is the task of a remedy for foot and mouth disease.[1]

Coffee has a strong stimulating effect on the brain and on the senses. It is also a powerful regulator of the metabolism. It was prepared in a special way for use in the treatment of FMD. It was applied to the animal intravenously. This is fully reported in *Agriculture of Tomorrow*. Although many reasonably successful trials were undertaken, the treatment never really caught on because as a notifiable disease FMD was considered untreatable.

In 2001, at the height of the epidemic, all farmers had to observe strict biosecurity rules to prevent animals becoming infected. A policy for the complete eradication of the disease was in place and ever more draconian measures were introduced, including a policy of creating so-called 'fire breaks' around infected farms. This meant culling healthy animals within a three-mile radius of an infected farm. It was this cruel policy that caused a rebellion among farmers. The biodynamic farm of Oaklands Park in Newnham, Gloucestershire, came under threat when a neighbouring farm 'tested positive' to FMD. Men from the ministry backed by the army were to be sent in the next day to slaughter the farm's milking herd. That it did not happen was due to the courage and commitment of everyone in the farm community and the hundreds of supporters from across the county who came early in the morning and literally prevented access to the farm by barring the road. The cull attempt was then called off and this marked a turning point in the battle for a more humane approach.

The response to the foot-and-mouth epidemic is, of course, just one of many major battles that supporters of a sustainable, ecological approach to the earth have had to fight at the end of the twentieth and the beginning of the twenty-first centuries.

Some years before the FMD outbreak another major and previously unknown animal disease had broken out in Britain. Bovine Spongiform Encephalitis (BSE) – also known as Mad Cow Disease – affected the brain and caused a form of madness in cattle. It was traced back to the feeding of meat residues to these herbivores and certain strong chemicals used in the treatment of warble fly, although the latter was never openly admitted by the ministry. Interestingly, Rudolf Steiner once remarked that if cows were required to eat meat they would go crazy.[2]

The response of the authorities to this disease, however, presented a particular challenge for the biodynamic movement. Since the animal's nervous system is affected by the disease there was a concern that the careless disposal of animal by-products containing nerve tissue might be a threat to human health. Butchers and abattoirs had therefore to ensure that all suspect animal parts were disposed of safely and could not enter the food chain. This meant that brain material, which effectively means the heads of cattle, sheep and goats, was deemed to be 'high risk material' along with other organs such as intestines and mesentery, which are also penetrated with nerves. Since these are the organs that, along with the horns of cows, are needed for the biodynamic preparations, the policy became a serious problem. Concerted attempts were made across Europe to engage with the authorities and many discussions took place with the European Commission and with the Ministry of Agriculture, as it then was, in the UK. There was some success but most of the restrictions remained. Today, although some restrictions are being relaxed across Europe, the obtaining of these organs remains a problem, particularly in the UK.

Another big issue that came to the fore in the last years of the twentieth century was that of the genetic modification of food. The biodynamic movement joined forces with the wider organic movement to oppose the use of all GM seeds and exclude them from organic systems. The public campaign grew, became widespread and ultimately prevented the entire food chain from being contaminated with GMOs in 1998. A key moment was the publication by *Ecologist* magazine of a detailed account of the corporate biography of Monsanto, the leading purveyor of GMOs at the time. In describing how the company had been operating over many decades – ruthlessly pursuing profit and producing agri-chemicals, including the infamous 'agent orange' used in Vietnam – it made visible what had been hidden from public view. From that moment on, the GMO drive lost traction and it has taken a long time to again become a global threat. Today, two decades later, we are facing the imposition of what are now called gene-edited crops. These differ from GMOs solely in the fact that instead of genes from another species being used, it is only the plant's own genes that are changed or edited.

The biodynamic approach always begins by considering the whole organism and its relationship to the environment. Although what is inherited can be localised in the genes, the environment, soil and cultivation have a longer-term effect on the characteristics of a plant. Biodynamic farmers understand that there is no need to interfere with the integrity of the plant cell.

13. Towards a Biodynamic Future

The biodynamic movement is now approaching its centenary. During the course of its first one hundred years, the seeds planted in Eastern Europe have germinated and grown into a diverse worldwide movement. The inspiration given at Koberwitz offers universally applicable ideas for understanding nature and agricultural processes. The human being is the archetype and model for the biodynamic farm, and its further evolution, like evolution as a whole, is a journey towards ever-greater individualisation. It is the biodynamic farmer's task to transform the living organism of the farm into an individuality, and there are as many ways of doing this as there are farmers.

The biodynamic movement, however, is also a global movement. Can we individualise and globalise at the same time? The experience of the biodynamic movement suggests that this is indeed possible. For example, a research project was recently carried out by the Agriculture Section at the Goetheanum to assess the different ways in which the biodynamic preparations are made and applied across the world.[1] This project confirmed that while the basic principles involved remain constant there are a wide range of working practices and approaches to the preparations. Each one is experienced as being right and valid for the specific people, places and conditions where it is applied. This demonstrates the living nature of the activity and shows how individual human interest and engagement is more important than externally imposed procedures and quality criteria. It is also an example of how universal principles can be tailored to the specific requirements of the individual locality.

Human beings have been transforming the earth ever since farming began. Nature has been tamed, crops cultivated and animals domesticated. There has been an enormous amount of environmental destruction going on during that time too. There is growing awareness of the need to care for our planet, acknowledge the value of the world's ecosystems and engage with organic and nature-friendly practices. There is, however, a strong movement towards ending agriculture and rewilding large areas of the world in the name of tackling climate change. There are even ideas of restricting humanity to half of the planet.[2] A landscape abandoned and left entirely to itself, however, has no future, nor can the wilderness of old be recreated. Only human engagement can bring renewal about. A farmer is needed to cultivate and form the landscape. This means improving the fertility of the soil beyond what naturally occurs and producing abundant food of high quality, while at the same time ensuring that complex inter-relationships and the balance of nature are maintained. It also means artistically and sensitively creating areas of wildness in the landscape that can support the organism of the farm and its surroundings. Biodynamic agriculture in this way holds the key to renewing and maintaining life on the planet.

Another issue through which biodynamic agriculture needs to steer a careful path is climate change. Changes are undoubtedly happening and many are caused by human civilisation, deforestation and poor husbandry practices, and it is these that we need to focus on.

It is also widely stated that the earth is overpopulated and that there is insufficient food to go around. In reality there is more than enough for everyone; the problem is essentially one of poor distribution and corporate control of the food chain. If biodynamic techniques were used to farm the land all over the world and production was carried out on a small-scale to serve local communities, there would be an ample supply. This is being recognised in many parts of the world, such as in India where thousands of smallholders are using biodynamic methods to grow their food. Indeed the enthusiasm for the biodynamic approach in India and elsewhere in the south, far exceeds that in many parts of Europe.

It will be ever more important as time goes on to stand by what is essential for biodynamic agriculture and continue to resist measures that undermine it, for example, gene-edited and hybrid seeds, and applications of technology that impact on life processes, such as micro-wave radiation, and also those which in the interest of efficiency and convenience sever the relationship between the human being and nature.

It is also vital that the movement doesn't compromise its fundamental principles. Very early on in the development of the movement an attempt was made to break the link between the new agriculture and its esoteric foundations in anthroposophy. During a conference held at the Marienhöhe Farm near Berlin in July 1928, two weeks before the World Conference on Spiritual Science in London, the newly formed German Experimental Circle agreed to separate the new agricultural methods from their anthroposophical source and focus on a practical approach.[3] This was a questionable though understandable step to take in view of the growing interest in biodynamic agriculture from farmers in the wider farming community. The connection to the source was nevertheless maintained by committed biodynamic farmers. The founding of the BDA in the UK in the 1930s was another attempt to separate biodynamic agriculture from its source, but thanks to the commitment of many this was also overcome and the connection has remained. This is important since the vitality of the movement depends on retaining its connection to the spiritual-scientific impulse of Rudolf Steiner. It is especially important today when considering the trend towards the so-called 'conventionalisation' of biodynamic and organic agriculture, particularly where large-scale farming operations are concerned.

The journey of biodynamic agriculture – from its earliest beginnings in Silesia, through the struggles of the twentieth century to the time in which we are now living – has shown how versatile and adaptable it can be. Biodynamic food and a real-life connection to nature, the earth and the spiritual world will become ever more important as we face the dehumanising tendencies of materialism that confront us today.

Appendix 1: Office Holders of the Anthropsophical Agricultural Foundation (AAF)

AGM Date	Chair/ President	Secretary (S)/ Treasurer (T)	Council	Fieldsman (F)/ Editor (E)
Nov 1928.[1]	D. Dunlop	M. Pease (S)	G. Kaufmann, M. Wood.	C. Mirbt (C. Mier) (F)
Dec 1930.	D. Dunlop	M. Pease (S), H. Lloyd-Wilson (T).	G. Kaufmann, M. Wood, Lady Mackinnon, E. Merry, M. Cameron, M. Cross.	C. Mirbt (F)
Dec 1931.[2]	D. Dunlop	M. Pease (S), H. Lloyd-Wilson (T).	G. Kaufmann, M. Wood, Lady Mackinnon, E. Merry, M. Cameron, M. Cross.	C. Mirbt (F) – to Stetchford.
Dec 3, 1932. RSH, London.	D. Dunlop/ G. Wachsmuth.	M. Pease (S), H. Lloyd-Wilson (T).	G. Kaufmann, M. Wood, E. Merry, M. Cameron, M. Cross, G. Mirbt, L. L. Binnie.	C. Mirbt (F)
Nov 25, 1933. RSH, London.	D. Dunlop/ G. Wachsmuth.	M. Pease (S), H. Lloyd-Wilson (T).	G. Kaufmann, M. Wood, E. Merry, M. Cameron, M. Cross, G. Mirbt, L. L. Binnie, M. Bruce.	C. Mirbt (F)

AGM Date	Chair/ President	Secretary (S)/ Treasurer (T)	Council	Fieldsman (F)/ Editor (E)
Nov 24, 1934. RSH, London.[3]	D. Dunlop	M. Pease (S), H. Lloyd-Wilson (T).	G. Kaufmann, M. Wood, E. Merry, M. Cameron, G. Mirbt, M. Bruce.	C. Mirbt (F) – to Clent.
Oct 5, 1935. RSH, London.[4]	G. Kaufmann	M. Pease (S), H. Lloyd-Wilson (T).	G. Kaufmann, M. Wood, E. Merry, M. Cameron, G. Mirbt, M. Bruce, F. Yuille-Smith, C. B. Davy.	C. Mirbt (F)
Oct 25, 1936. RSH, London.	G. Kaufmann	M. Pease (S), M. Wheeler (T).	M. Bruce, M. Cameron, Dr E. Kolisko, E. Merry, G. Mirbt, M. Wood, F. Yuille-Smith.	C. Mirbt (F)
Oct 23, 1937. RSH, London.[5]	G. Kaufmann	M. Pease (S), O. Mainland.	M. Cameron, C. Davy, M. Edwards, Dr E. Kolisko, E. Merry, G. Mirbt, M. Wood, F. Yuille-Smith.	C. Mirbt (F)
Oct 22, 1938. RSH, London.	G. Kaufmann	M. Pease (S), O. Mainland (T).	M. Cameron, C. Davy, M. Edwards, Dr E. Kolisko, E. Merry, M. Wood, F. Yuille-Smith.	C. Mirbt (F)

AGM Date	Chair/ President	Secretary (S)/ Treasurer (T)	Council	Fieldsman (F)/ Editor (E)
Nov 24, 1934. RSH, London.[3]	D. Dunlop	M. Pease (S), H. Lloyd-Wilson (T).	G. Kaufmann, M. Wood, E. Merry, M. Cameron, G. Mirbt, M. Bruce.	C. Mirbt (F) – to Clent.
Oct 5, 1935. RSH, London.[4]	G. Kaufmann	M. Pease (S), H. Lloyd-Wilson (T).	G. Kaufmann, M. Wood, E. Merry, M. Cameron, G. Mirbt, M. Bruce, F. Yuille-Smith, C. B. Davy.	C. Mirbt (F)
Oct 25, 1936. RSH, London.	G. Kaufmann	M. Pease (S), M. Wheeler (T).	M. Bruce, M. Cameron, Dr E. Kolisko, E. Merry, G. Mirbt, M. Wood, F. Yuille-Smith.	C. Mirbt (F)
Oct 23, 1937. RSH, London.[5]	G. Kaufmann	M. Pease (S), O. Mainland.	M. Cameron, C. Davy, M. Edwards, Dr E. Kolisko, E. Merry, G. Mirbt, M. Wood, F. Yuille-Smith.	C. Mirbt (F)
Oct 22, 1938. RSH, London.	G. Kaufmann	M. Pease (S), O. Mainland (T).	M. Cameron, C. Davy, M. Edwards, Dr E. Kolisko, E. Merry, M. Wood, F. Yuille-Smith.	C. Mirbt (F)

AGM Date	Chair/ President	Secretary (S)/ Treasurer (T)	Council	Fieldsman (F)/ Editor (E)
1939	G. Kaufmann	M. Pease (S), O. Mainland (T).	No AGM 1939–44.	C. Mirbt (F), D. Davy (E).
1940	G. Kaufmann[6]	M. Pease (S), O. Mainland (T).	—	C. Mirbt,[7] D. Davy (E).
1945	G. Adams	M. Pease (S),[8] O. Mainland (T).	—	C. Mier (F), D. Davy (E).
1946	D. Clement	C. Mier (S. & T)	G. Adams, K. Brocklebank, M. Cameron, D. Davy, M. Edwards, J. Eliot, F. Lambe, G. Mier, O. Whicher, M. Wood.	C. Mier (F), D. Davy (E).
1947	D. Clement	C. Mier (S), K. Brocklebank (T).	—	C. Mier (F), D. Davy (E).
1948	D. Clement	C. Mier (S), K. Brocklebank (T).	—	C. Mier (F), D. Davy (E).
1949	D. Clement	C. Mier (S), K. Brocklebank (T).	G. Adams, M. Edwards, H. Ellis, F. Lambe, G. Mier, Nicholls, O. Whicher, M. Wood.	C. Mier (F), D. Davy (E).
1950	D. Clement	C. Mier (S), K. Brocklebank (T).	G. Adams, M. Edwards, H. Ellis, F. Lambe, G. Mier, Nicholls, O. Whicher, M. Wood.	C. Mier (F), D. Davy (E).

Appendix 2: Office Holders of the Biodynamic Association (BDA)

The Biodynamic Association (BDA) was founded in the spring of 1937 following the split in the Anthroposophical Society.

AGM Date	President/ Vice-President (VP)	Chair	Secretary (S)/ Treasurer (T)/ Editor (E)	Council
Jun 11, 1938.	Lady Merthyr (P), H. Collison (VP), J. Ferguson (VP).	Lady McKinnon	H. Popplebaum (S & T), M. Cross (E).	G. Bacchus, B. Saunders-Davies.
Jun 10, 1939.[1]	Lady Merthyr (P), H. Collison (VP), J. Ferguson (VP).	Lady McKinnon	H. Popplebaum (S & T), M. Cross (E).	G. Bacchus, B. Saunders-Davies.
Apr 20, 1940.[2]	J. Ferguson (P), H. Collison (VP).	Lady McKinnon	M. Murchison (S), H. Popplebaum (T), M. Cross (E).	G. Bacchus, B. Saunders-Davies.
Jun 22, 1943.	E. Pfeiffer (P), H. Collison (VP).	J. Ferguson	M. Murchison (S), Lady McKinnon (T), M. Cross (E).	G. Bacchus, R. Gardiner, E. Murray-Usher.
Jun 14, 1945.	E. Pfeiffer (P), H. Collison (VP).	J. Ferguson	M. Murchison (S), Lady McKinnon (T), M. Cross (E).	R. Gardiner, E. Murray-Usher.
May 13, 1946.	E. Pfeiffer (P), Lady McKinnon (VP).	J. Ferguson	Lady McKinnon (S & T), M. Cross (E).	N. Fisher, R. Gardiner, E. Murray-Usher.

AGM Date	President/ Vice-President (VP)	Chair	Secretary (S)/ Treasurer (T)/ Editor (E)	Council
May 19, 1947.	E. Pfeiffer (P), Lady McKinnon (VP).	J. Ferguson	K. Thornton (S), E. Drew (T), M. Cross (E).	N. Fisher, R. Gardiner, H. Maude, E. Murray-Usher, R. Raab, C. Savoury, P. Thomas.
May 19, 1948.	E. Pfeiffer (P), Lady McKinnon (VP).	J. Ferguson	K. Thornton (S), E. Drew (T), M. Cross (E).	N. Fisher, H. Maude, E. Murray-Usher, R. Raab, C. Savoury, P. Thomas.
Jun 16, 1949.	E. Pfeiffer (P), Lady McKinnon (VP).	R. Gardiner	K. Thornton (S), E. Drew (T), M. Cross (E).	N. Fisher, H. Maude, E. Murray-Usher, R. Raab, C. Savoury, P. Thomas.
Jun 13, 1950.	E. Pfeiffer (P)	R. Gardiner	K. Thornton (S), E. Drew (T), M. Cross (E).	N. Fisher, J. Jeffree, H. Maude, E. Murray-Usher, R. Raab, C. Savoury, P. H. Thomas, K. Thornton.

The Biodynamic Association (BDA) and the Anthroposophical Agricultural Foundation (AAF) unite to form the Biodynamic Agricultural Association (BDAA) in 1950. The newsletter of the BDA. becomes the new organisation's external organ of communication, and *Notes and Correspondence*, the internal organ.

Appendix 3: Office Holders of the Biodynamic Agricultural Association (BDAA)

The Biodynamic Association (BDA) and the Anthroposophical Agricultural Foundation (AAF) united to form the Biodynamic Agricultural Association (BDAA) in 1950. In 1953, *Notes and Correspondence* was relaunched as *Star & Furrow*.

Key to Executive Functions:

ED: Executive Director[1]
F: Fieldsman
N&C: Editor, *Notes and Correspondence*
S&F: Editor, *Star & Furrow*
SDO: Seed Development Officer[2]

AGM Date	Chair	Secretary (S)/ Treasurer (T)	Council	Executive Functions
1951. Rudolf Steiner House (RSH), London.	D. Clement	K. Brocklebank (S), E. Drew (T).	G. Adams, F. Lambe, O Whicher, H. Ellis, M. Wood, J. Jeffree, K. Thornton, Fisher, J. Nichols, M. Edwards, Maude, G. Mier, Thomas, R. Gardiner, Savory.[3]	D. Davy (F), M. Cross (N&C).
1952. RSH, London.	D. Clement	C. Mier (S), K. Brocklebank (T).	G. Adams, J. Jeffree, A. Kaye, B. Mansfield, O. Whicher, M. Wood, K. Thornton, H. Ellis.	D. Davy (F)
1953. RSH, London.	D. Clement	C. Mier (S), K. Brocklebank (T).	G. Adams, J. Jeffree, A. Kaye, B. Mansfield, O. Whicher, M. Wood, K. Thornton, H. Ellis.	(N&C changes to S&F.) C. Mier (F), D. Davy (S&F).
1954. RSH, London.	D. Clement	C. Mier (S), K. Brocklebank (T).	G. Adams, J. Jeffree, A. Kaye, O. Whicher, K. Brocklebank, M. Wood, K. Thornton, M. Geuter.	D. Davy (S&F)
1955. RSH, London.	D. Clement	(Office moves to Clent.) C. Mier (S), B. Mansfield (T).	G. Adams, J. Jeffries, A. Kaye, O. Whicher, K. Thornton, M. Wood, M. Geuter, D. Duffy, M. Millett, K. Brocklebank.	G. Corrin (F), D. Davy (S&F).
1956. RSH, London.	D. Clement	C. Chance[4] (S), B. Mansfield (T).	J. Jeffries, A. Kaye, O. Whicher, M. Wood, M. Geuter, M. Millett, K. Brocklebank, C. Mier, F. Lambe, D. Duffy.	G. Corrin (F), D. Davy (S&F).

AGM Date	Chair	Secretary (S)/ Treasurer (T)	Council	Executive Functions
1957. RSH, London.	D. Clement	C. Chance (S & T)	J. Jeffries, A. Kaye, O. Whicher, B. Mansfield, M. Wood, M. Geuter, M. Millett, K. Brocklebank, C. Mier, F. Lambe, D. Duffy, A. Hanbury-Sparrow.	G. Corrin (F), D. Davy (S&F).
1958. RSH, London.	D. Clement	C. Chance (S & T)	A. Kaye, B. Mansfield, M. Wood, M. Geuter, M. Millett, K. Brocklebank, C. Mier, F. Lambe, D. Duffy, A. Brockman, A. Henbury-Sparrow, M. Hirst, Mrs Somerwell.	G. Corrin (F), D. Davy (S&F).
1959. RSH, London.	D. Clement	C. Chance (S), J. Soper (T).	A. Kaye, B. Mansfield, M. Wood, M. Geuter, M. Millett, K. Brocklebank, C. Mier, A. Henbury-Sparrow, A. Brockman, W. Wells.	G. Corrin (F), D. Davy (S&F).
1960. RSH, London.	D. Clement	C. Chance (S), J. Soper (T).	A. Kaye, B. Mansfield, M. Wood, M. Geuter, M. Millet, K. Brocklebank A. Henbury-Sparrow, A. Brockman, S. Rudel, C. Mier, W. Wells.	G. Corrin (F), D. Davy (S&F).
Dec 9, 1961. RSH, London.	D. Clement	C. Chance (S), J. Soper (T).	A. Kaye, B. Mansfield, M. Geuter, K. Brocklebank, C. Mier, A. Brockman, S. Rudel, W. Wells, K. Castelliz, M. Edwards. A. Hanbury-Sparrow,	G. Corrin (F), D. Davy (S&F).

AGM Date	Chair	Secretary (S)/ Treasurer (T)	Council	Executive Functions
Dec 8, 1962. RSH, London.	D. Clement	C. Chance (S), J. Soper (T).	A. Kaye, B. Mansfield, K. Brocklebank, C. Mier, A. Hanbury-Sparrow, A. Brockman, S. Rudel, K. Castelliz, M. Edwards, W. Wells.	G. Corrin (F), D. Davy (S&F).
Dec 7, 1963. RSH, London.	D. Clement	J. Soper (S & T)	A. Kaye, B. Mansfield, S. Rudel, K. Brocklebank, C. Mier, A. Henbury-Sparrow, A. Brockman, K. Castelliz, M. Edwards, D. Munyard.	G. Corrin (F), D. Davy (S&F).
Dec 5, 1964. RSH, London.	D. Clement	J. Soper (S & T)	A. Kaye, K. Brocklebank, C. Mier, A. Hanbury-Sparrow, C. Obery, A. Brockman, S. Rudel, M. Geuter, K. Castelliz, M. Edwards.	G. Corrin (F), D. Davy (S&F).
Dec 4, 1965.[5] RSH, London.	D. Clement	J. Soper (S & T)	A. Kaye, K. Brocklebank C. Mier, A. Brockman, S. Rudel, K. Castelliz, M. Edwards, C. Obery.	G. Corrin (F), D. Davy (S&F).
1966. RSH, London.	D. Clement	J. Soper (S & T)	A. Kaye, A. Brockman, S. Rudel, K. Castelliz, M. Edwards, C. Obery, D. Munyard, O. Matthews.	G. Corrin (F), D. Davy (S&F).
Apr 6, 1967. RSH, London.	D. Clement	J. Soper (S & T)	A. Kaye, A. Brockman, S. Rudel, K. Castelliz, DC. Obery, D. Munyard, M. Edwards, O. Matthews.	G. Corrin (F), D. Davy (S&F).
Apr 8, 1968. RSH, London.	D. Clement	J. Soper (S & T)	A. Kaye, A. Brockman, S. Rudel, K. Castelliz, D. Obery, D. Munyard, M. Edwards, O. Matthews.	G. Corrin (F), D. Davy (S&F).

AGM Date	Chair	Secretary (S)/ Treasurer (T)	Council	Executive Functions
Dec 6, 1969. RSH, London.	D. Clement	J. Soper (S & T)	A. Kaye, A. Brockman, S. Rudel, K. Castelliz, D. Obery, M. Courtney, M. Soper, O. Matthews, M. Edwards, D. Munyard.	G. Corrin (F), D. Davy (S&F).
Dec 12, 1970. RSH, London.	D. Clement	J. Soper (S & T)	A. Kaye, A. Brockman, S. Rudel, K. Castelliz, D. Obery, M. Courtney.	G. Corrin (F), M. Soper (S&F).
Dec 4, 1971. RSH, London.	D. Clement	J. Soper (S & T)	A. Kaye, A. Brockman, S. Rudel, K. Castelliz, D. Obery, M. Courtney, H. Koepf.	G. Corrin (F), M. Soper (S&F).
Dec 9, 1972. RSH, London.	D. Clement	J. Soper (S & T)	A. Kaye, A. Brockman, S. Rudel, K. Castelliz, C. Obery, M. Courtney, H. Koepf.	G. Corrin (F), M. Soper (S&F).
Dec 8, 1973. RSH, London.	D. Clement	J. Soper (S & T)	A. Kaye, A. Brockman, S. Rudel, K. Castelliz, C. Obery, M. Courtney, H. Koepf, M. Schmundt.	G. Corrin (F), M. Soper (S&F).
Dec 7, 1974. RSH, London.	D. Clement	J. Soper (S & T)	K. Castelliz, A. Kaye, C. Obery, H. Koepf, S. Rudel. M. Schmundt, A. Brockman, M. Courtney.	G. Corrin (F), M. Soper (S&F).
Nov 29, 1975. RSH, London.	D. Clement	J. Soper (S & T)	A. Kaye, C. Obery, K. Castelliz, M. Schmundt, H. Koepf, S. Rudel, M. Courtney, A. Brockman.	G. Corrin (F), M. Soper (S&F).
Dec 4, 1976. RSH, London.	D. Clement	J. Soper (S & T)	A. Kaye, C. Obery, K. Castelliz, J. Anderson, C. Budd, G. Sheppard, B. Notrott, T. v Freeden, H. Koepf.	G. Corrin (F), M. Soper (S&F).

AGM Date	Chair	Secretary (S)/ Treasurer (T)	Council	Executive Functions
Dec 3, 1977. RSH, London.	D. Clement	J. Soper (S & T)	A. Kaye, M. Courtney, B. Nottrot, G. Shepperd, A. Brockman, C. Budd, K. Castelliz, J. Dorrell.	G. Corrin (F), M. Soper (S&F).
Dec 9, 1978. RSH, London.	D. Clement	J. Soper (S & T)	M. Courtney, G. Shepperd, A. Kaye, A. Brockman, C. Budd, K. Castelliz, J. Dorrell, B. Saunders-Davies.	G. Corrin (F), M. Soper (S&F).
Dec 8, 1979. RSH, London.	D. Clement	H. Fynes-Clinton (S), M. Wood (T).	H. Koepf, A. Kaye, J. Anderson, A. Brockman, M. Courtney, J. Dorrel, G. Shepperd, M. Heasman, K. Castelliz, B. Saunders-Davies.	G. Corrin (F), M. Soper (S&F).
Dec 6, 1980. RSH, London.	D. Clement	H. Fynes-Clinton (S), M. Wood (T).	A. Kaye, J. Soper, J. Anderson, A. Brockman, M. Courtney, J. Dorrel, G. Shepperd, C. Budd, K. Castelliz, M. Heasman, T. v Freeden, B. Saunders-Davies.	G. Corrin (F), M. Soper (S&F).
Dec 5, 1981. RSH, London.	D. Clement	H. Fynes-Clinton (S), M. Wood (T).	A. Kaye, J. Soper, J. Anderson, A. Brockman, M. Courtney, J. Dorrel, G. Shepperd, C. Budd, M. Heasman, T. v Freeden, B. Saunders-Davies.	G. Corrin (F), M. Soper (S&F).
Dec 4, 1982. RSH, London.	D. Clement[6]	H. Fynes-Clinton (S), M. Wood (T).	A. Kaye, J. Soper, J. Anderson, A. Brockman, K. Castelliz, M. Courtney, J. Dorrel, G. Shepperd, R. Clarke, S. Naumann, B. Saunders-Davies, T. Clement, H. Koepf, C. Budd.	G. Corrin (F), O. Holbeck and B. Mansfield (S&F).

AGM Date	Chair	Secretary (S)/ Treasurer (T)	Council	Executive Functions
Dec 3, 1983. RSH, London.	D. Clement	H. Fynes-Clinton (S), M. Wood (T).	B. Saunders-Davies, T. Clement, J. Dorrel, G. Shepperd, R. Clarke, S. Naumann, W. Rudert, A. Kaye, J. Soper, M. Guepin, H. Koepf, K. Castelliz.	G. Corrin (F), O. Holbeck (S&F).
Dec 1, 1984. RSH, London.	D. Clement	H. Fynes-Clinton (S), M. Wood (T).	B. Saunders-Davies, T. Clement, J. Dorrel, G. Shepperd, R. Clarke, A. Kaye, S. Naumann, W. Rudert, J. Soper, M. Courtney.	G. Corrin (F), O. Holbeck (S&F).
1985. RSH, London.	D. Clement	D. Adams (S), M. Wood (T).	W. Rudert, T. Clement, A. Brockman, B. Jarman, J. Bradley, M. Heasman, M. Guepin, H. Koepf.	G. Corrin (F), O. Holbeck (S&F).
1986. RSH, London.	D. Clement	D. Adams (S), M. Wood (T).	A. Kaye, K. Castelliz, H. Koepf, A. Brockman, J. Anderson, M. Guepin, P. Thompson, W. Rudert, M. Newton, M. Wood, T. Clement, M. Heasman, B. Jarman, R. Evans, J. Bradley, J. Kirk.	G. Corrin (F), O. Holbeck (S&F).
1987. RSH, London.	D. Clement	D. Adams (S), R. Savage (T).	R. Thornton-Smith, E. Jenks, M. Schmundt, T. Clement, P. Thompson, A. Brockman, R. Evans, J. Grundmann.	O. Holbeck (S&F)
1988. RSH, London.	D. Clement	M. Jackson (S), R. Savage (T), P. Thompson (Event Secretary).	B. Jarman, P. Anderson, A. Kaye, M. Schmundt, R. Evans, R. Thornton-Smith, K. Castelliz, J. Twine, T. Matthews, H. Koepf, J. Grundmann, M. Newton, A. Brockman, E. Jenks.	O. Holbeck (S&F)

AGM Date	Chair	Secretary (S)/ Treasurer (T)	Council	Executive Functions
Dec 2, 1989. RSH, London.	J. Grundmann	A. Parsons (S), P. Thompson (Event Secretary).	A. Carnegie, T. Clement, J. Twine, B. Jarman, P. Anderson, K. Castelliz, R. Evans, T. Matthews, B. Saunders-Davies, M. Schmundt, R. Thornton-Smith, M. Newton.	P. Thompson (S&F)
Dec 1, 1990. RSH, London.	J. Grundmann / M. Newton.	A. Parsons (S), C. Reilly (T).	P. Anderson, K. Castelliz, R. Evans, B. Jarman, A. Kaye, H Koepf, T. Mathews, B. Saunders-Davies, M. Schmundt, R. Thornton-Smith, J. Pyzer, G. Sheppard, M. Newton.	P. Thompson (S&F)
Dec 7, 1991. Stroud.	T. v Voort	A. Parsons (S), C. Reilly (T).	P. Anderson, A. Brockman, M. Newton, K. Castelliz, R. Evans, B. Jarman, P. v Midden, J. Pyzer, B. Saunders-Davies, M. Schmundt, G. Sheppard, C. Wannop, T. Matthews.	P. Thompson (S&F)
1992. Ilkeston.	A. Brockman	A. Parsons (S), C. Reilly (T).	A. Blom, S. Bradley, T. Matthews, J. Carver, J. Pyzer, P. v Midden, B. Saunders-Davies, G. Sheppard, C. Wannop, R. Evans, K. Castelliz.	P. Thompson (S&F)
1993. Michael Hall, Forest Row.	A. Brockman	A. Parsons (S), C. Reilly (T).	A. Blom, S. Bradley, H. Mackay, A. Mathews, J. Pyzer, J. Twine, C. Wannop, B. Saunders-Davies, K. Castelliz, J. Brinch, R. Finnigan, R. Thornton-Smith, E. Wennekes.	P. Thompson (S&F)

AGM Date	Chair	Secretary (S)/ Treasurer (T)	Council	Executive Functions
1994. Edinburgh.	A. Brockman	A. Parsons (S), C. Reilly (T).	J. Brinch, P. Anderson, K. Castelliz, B. Saunders-Davies, A. Blom, S. Bradley, R. Finnegan, A. Mathews, R. Thornton-Smith, J. Twine, E. Wennekes, H. Mackay.	P. Thompson (S&F)
Dec 9, 1995. Hawkwood, Gloucester-shire.	B. Jarman	J. Paterson (S), H. Mackay (T), C. Wannop (Promotions).	J. Brinch, R. Thornton-Smith, J. Twine, E. Wennekes, S. Whipple. B. Saunders-Davies, K. Castelliz, P. Bateman, J. Kirk, F. Schikorr, P. Martin.	J. Anderson (F), P. Thompson (S&F).
Nov 23, 1996. Hereford.	B. Jarman	J. Paterson (S), H. Mackay (T).	P. Bateman, J. Brinch, J. Kirk, P. Martin, F. Schikorr, R. Thornton-Smith, S. Whipple, B. Saunders-Davies.	J. Anderson (F), P. Thompson (S&F).
Nov 22, 1997. Kings Langley.	B. Jarman	J. Paterson (S), I. Bailey (T).	J. Kirk, P. Martin, F. Schikorr, S. Whipple, P. Bateman, B. Saunders-Davies, A. Kaye, M. Bate, S. Berry, A. Irwin, V. Griffiths, S. Hall, P. v Midden.	J. Anderson (F), P. Thompson (S&F).
Oct 3, 1998. Stroud.	B. Jarman[7]	C. Eyles (S), I. Bailey (T).	V. Griffiths, S. Hall, A. Irwin, M. Bate, S. Berry, B. Saunders-Davies, P. Pieterse, M. Bate, W. Rudert.	J. Anderson (F), P. Thompson (S&F).
Oct 8, 1999. Trigonos, Caernarfon.	B. Jarman	J. Standing (S), I. Bailey (T).	V. Griffiths, A. Irwin, P. Pieterse, R. Swann, B. Saunders-Davies, M. Bate, P. Waller.	J. Anderson (F), P. Thompson (S&F).

AGM Date	Chair	Secretary (S)/ Treasurer (T)	Council	Executive Functions
2000. Newton Dee.	A. Irwin	J. Standing (S), I. Bailey (T).	A. Brockman, M. Bate, V. Griffiths, J. Miller, P. Waller, P. Pieterse.	B. Jarman (ED), J. Anderson (F), R. Swann (S&F).
Oct 13, 2001. Otley College, Ipswich.	A. Irwin	J. Standing (S), I. Bailey (T).	J. Miller, B. Saunders-Davies, P. Pieterse, P. Waller, C. Stockdale, N. Raeside.	B. Jarman (ED), P. Brinch (SDO), R. Swann (S&F).
Oct 5, 2002. Emerson College, Forest Row.	A. Irwin	J. Standing (S), I. Bailey (T).	J. Millar, N. Raeside, C. Stockdale, P. Waller, M. Bate, V. Griffiths.	B. Jarman (ED), P. Brinch (SDO), R. Swann (S&F).
Oct 4, 2003. Devon.	A. Irwin	J. Standing (S), I. Bailey (T).	M. Bate, V. Griffiths, J. Miller, N. Raeside, C. Stockdale, P. Waller, L. Dungworth, P. Fleming.	B. Jarman (ED), P. Brinch (SDO), R. Swann (S&F).
Oct 9, 2004. Norwich, Norfolk.	A. Irwin	J. Standing (S), I. Bailey (T).	M. Bate, L. Dungworth, V. Griffiths, N. Raeside, C. Stockdale, P. Fleming, P. v Vliet, L. Ellis.	B. Jarman (ED), P. Brinch (SDO), R. Swann (S&F).
Oct 8, 2005. Loch Arthur, Dumfries.	N. Raeside	J. Standing (S), I. Bailey (T).	L. Ellis, P. Fleming, V. Griffiths, A. Irwin, C. Stockdale, P. v Vliet, L. Dungworth.	B. Jarman (ED), P. Brinch (SDO), R. Swann (S&F).

AGM Date	Chair	Secretary (S)/ Treasurer (T)	Council	Executive Functions
Oct 7, 2006. Sturts Farm, Ringwood.	N. Raeside	J. Standing (S), I. Bailey (T).	R. Gantlet, L. Ellis, P. Fleming, C. Stockdale, P. v Vliet, L. Dungworth, C. Volmer.	B. Jarman (ED), R. Swann (S&F).
Oct 6, 2007. Borough Market, London.	N. Raeside	J. Standing (S), I. Bailey (T).	L. Ellis, R. Gantlet, T. Matthews, B. Cavendish, S. Parsons, T. Matthews, R. Thornton-Smith, R. Lord.	B. Jarman (ED), R. Swann (S&F).
Oct 11, 2008. Clanabogan, N. Ireland.	S. Parsons	J. Standing (S), I. Bailey (T).	B. Cavendish, R. Gantlet, T. Matthews, N. Raeside, R. Thornton-Smith, R. Lord.	B. Jarman (ED), R. Swann (S&F).
Oct 4, 2009. Laverstoke Farm, Basingstoke.	S. Parsons	J. Standing (S), I. Bailey (T).	P. Schofield, P. Brown, B. Cavendish, R. Lord, R. Thornton-Smith.	B. Jarman (ED), R. Swann (S&F).
Oct 9, 2010. BDAG,[8] Emerson College, Forest Row.	S. Parsons	J. Standing (S), I. Bailey (T).	R. Lord, P. Schofield, P. Brown, V. Griffiths, D. McGregor, C. v Bulow, H. Hermann, C. Stockdale, E. Austin.	P. Brown (ED), R. Swann (S&F).
Oct 8, 2011. Ruskin Mill, Nailsworth.	P. Brown	J. Standing (S), I. Bailey (T).	V. Griffiths, C. v Bulow, H. Herrmann, C. Stockdale, E. Austin, S. Christy, J. Lister.	T. Brink (ED), R. Swann (S&F).
Oct 6, 2012. Garden Organic, Ryton.	P. Brown	J. Standing (S), I. Bailey (T).	V. Griffiths, C. von Bulow, H. Hermann, C. Stockdale, E. Austin, S. Christy, J. Lister.	S. Parsons (ED), R. Swann (S&F).

AGM Date	Chair	Secretary (S)/ Treasurer (T)	Council	Executive Functions
Oct 12, 2013. Brantwood Cumbria.	P. Brown	J. Standing (S), E. Holmes (T).	E. Austin, S. Christy, J. Lister, K. Lange, L. Cammish, H. Herrmann.	P. Brown (ED), R. Swann (S&F).
Oct 4, 2014. Steiner Academy, Hereford.	C. Stockdale	J. Standing (S), E. Holmes (T).	R. O'Kelly, J. Lister, E. Austin, H. Herrmann, R. O'Kelly, G. John, S. Christy, I. Bailey.	P. Brown (ED), R. Swann (S&F).
Oct 5, 2015. Hood Manor, Devon.	C. Stockdale	J. Standing (S), E. Holmes (T).	J. Lister, S. Christy, L. Cammish, R. O'Kelly, G. John, S. Christopher-Bowes, I. Bailey.	P. Brown (ED), R. Swann (S&F).
Oct 8, 2016. Lauriston Farm, Essex.	C. Stockdale	J. Standing (S), E. Holmes (T).	J. Lister, G. John, S. Christopher-Bowes, L. Ellis, B. R. O'Kelly, B. Krehl, L. Brown, I. Bailey.	P. Brown (ED), R. Swann (S&F).
Oct 7, 2017. Tablehurst Farm, Forest Row.	C. Stockdale	J. Standing (S), E. Holmes (T).	S. Christopher-Bowes, L. Ellis, B. Krehl, L. Brown, S. Christy, O. Kirst, J. Williams, J. Lister.	L. Brown (ED), R. Swann (S&F).
Oct 6, 2018. Aura Soma, Lincoln-shire.	J. Wright	J. Standing (S), A. Drevson.	L. Ellis, B. Krehl, L. Brown, S. Christy, O. Kirst, J. Lister, P. Brown, C. Stockdale, H. Steenbergen, R. Romano, B. Bowmaker, D. Morris.	L. Brown (ED), R. Swann (S&F).
Oct 5, 2019. Glasshouse College, Stour-bridge.	J. Wright	J. Standing (S), A. Drevson.	S. Christy, O. Kirst, P. Brown, C. Stockdale, H. Steenbergen, R. Romano B. Bowmaker, I. Bailey, A. Tranquelini, J. Couling.	G. Kaye (ED), R. Swann (S&F).

AGM Date	Chair	Secretary (S)/ Treasurer (T)	Council	Executive Functions
Oct 3, 2020. Online.	J. Wright	J. Standing (S), S. Christy (T).	P. Brown, C. Stockdale, H. Steenbergen, R. Romano, D. Morris, I. Bailey, A. Tranquellini, J. Couling, A. Drevson, J. Rosenbrock, M-L. Nukis, O. Kirst.	G. Kaye (ED), R. Swann (S&F).
Oct 16, 2021. Online.	J. Wright	J. Standing (S), S. Christy (T).	H. Steenbergen, I. Bailey, A. Tranquelini, A. Drevson, J. Rosenbrock, J. Foster, H. Robson, T. Martyn, A. Scott, O. Kirst, E. Cheong.	G. Kaye (ED), R. Swann (S&F).
Jan 14, 2022. Ruskin Mill, Nailsworth.	J. Rosenbrock	J. Standing (S), S. Christy (T).	A. Tranquelini, J. Rosenbrock, M-L. Nukis, J. Foster, H. Robson, T. Martyn, A. Scott, L. Findlay, I. Thomas, I. Bailey, O. Kirst.	G. Kaye (ED), R. Swann (S&F).
Oct 14, 2023. Ruskin Mill, Nailsworth.	J. Rosenbrock	J. Standing (S), S. Christy (T).	I. Bailey, T. Petherick, R. Thornton-Smith.	G. Kaye (ED), R. Swann (S&F).

Acknowledgements

My thanks to the following people for their support and encouragement: Vivian Griffiths, who provided invaluable insights into the history and development of the biodynamic movement in Britain; Jessica Standing and Gabriel Kaye from the Biodynamic Association and Mark Moodie from Considera who, along with many others, helped me access archive material; and Christian Maclean, whose original suggestion inspired me to write this book.

I would also like to acknowledge the debt I owe to Trauger Groh, who first taught me biodynamic farming and the value of community, and to the many other friends and farmers throughout the movement who have influenced my life.

Notes

Introduction

1. Steiner, *Agriculture Course*, p. 23.
2. Goethe, Johann Wolfgang von, 'Analysis and Synthesis', in Naydler, Jeremy (ed.), *Goethe on Science: An Anthology of Goethe's Scientific Writings*, Floris Books, UK 1996, p. 59.

1. The Beginning

1. Eugen von Keyserlingk left his papers and spider collection to the Natural History Museum in London.
2. *The Birth of a New Agriculture*, published by Temple Lodge Press, is a collection of articles and impressions about the Koberwitz course, edited by Count Adalbert Graf von Keyserlingk.
3. Ibid., p.48.
4. See The Address, June 11, 1924, in the *Agriculture Course.*
5. Mier, Gertrude, 'Life at Koberwitz', *Star and Furrow*, No, 43, Autumn 1974.
6. Sometime around 1945, Carl Mier changed his name from Carl Mirbt to make it easier to pronounce in English.

2. The Agriculture Course and the Experimental Circle

1. Steiner, *Agriculture Course*, p. 64.
2. Wachsmuth, Günther 'Agricultural Work – on the lines indicated by Rudolf Steiner', *Notes and Correspondence* Vol III, no. 1, March 1931, p.19.
3. Stapping, Walter, *Die Düngerpräparate Rudolf Steiners* [The Fertilizer Preparations by Rudolf Steiner], p. 456.
4. Steiner, *Agriculture Course*, p. 64.
5. In *Notes and Correspondence*, reporting on the founding of the AAF.
6. Author interview with Alan Brockman.
7. David Clement, speaking at a meeting of the Experimental Circle 2002.
8. Castelliz, Katherine, 'The Experimental Circle Past, Present and Future', *Star and Furrow* 1995.

3. Daniel Dunlop and the Anthroposophical Agricultural Foundation

1. See Rudolf Steiner, *Initiation Science and the Development of the Human Mind* (CW228), Rudolf Steiner Press, UK 2016. See in particular the lecture of September 10, 1923, 'The Druid Priest's Sun Initiation and His Perception of the Moon Spirits'.
2. Mier, Carl, 'Agriculture in the Present Age', in *Notes and Correspondence*, August 1937.
3. Image taken from John Paull, 'The Pioneers of Biodynamics in Great Britain: From Anthroposophic Farming to Organic Agriculture (1924–1940)' in *Journal of Environment Protection and Sustainable Development*, 5(4), 2019, pp. 138-145.
4. Mainland, Olive, 'Reminiscences of the Old Mill House', *Notes and Correspondence*, Autumn 1950.
5. Ibid.

4. The Challenge of the 1930s

1. Ita Wegman was a medical doctor and had been appointed leader of the Medical Section. Elizabeth Vreede was head of the Astronomical-Mathematical Section.

5. The Early Discoveries

1. Published in *Notes and Correspondences*, 1935.
2. See www.koliskoinstitute.org.
3. Kolisko, Lili, *Physiologischer und Physischer Nachweis der Wirksamkeit kleinster Entitäten* [Physiological and Physical Proof of the Efficacy of the Smallest Entities], Der Kommende tag Verlag, Stuttgart 1923.
4. Steiner, *The Evolution of Consciousness*, p. 249.
5. Knapp, G.A., 'In Memorium', *Star & Furrow*, No. 48 (1977).
6. Kolisko, Eugen and Lili, *Agriculture of Tomorrow*. See Chapter 2: Moon and Plant Growth.
7. Kolisko, Lili, *Eugen Kolisko – ein Lebensbild* [Eugen Kolisko: A Biography], Hohenloher Druck und Verlagshaus, Germany 1961.
8. Kolisko, Lili, *Agriculture of Tomorrow*, Chapter 2: Moon and Plant Growth.
9. Ibid.

6. The Organic Midwife

1. Paull, John, 'The Betteshanger Summer School: Missing Link between Biodynamic and Organic Farming', ResearchGate, 2014.
2. David Clement, *Experimental Circle*, 2002.
3. Image taken from John Paull, 'Lord Northbourne, the Man Who Invented Organic Farming: a Biography', in *Journal of Organic Systems*, 9(1), 2014, pp. 31-35.

4. Northbourne, Lord, *Of the Land and Spirit*, World Wisdom, USA 2008.
5. Paull, John, 'The Betteshanger Summer School: Missing Link between Biodynamic and Organic Farming', p. 23.
6. An excellent book documenting such ancient practices is Franklin C. King's *Farmers of Forty Centuries*. King travelled to China in around 1907 where he recorded how farmers made compost and improved the fertility of the land, how they laboriously dredged canals and used the mud layered with green matter to fertilise the land. He also described many other practices that helped to ensure the growing lands of China retained their fertility.
7. Northbourne, Lord, *Of the Land and Spirit*, p. 6.
8. Report in *Craven Herald & Pioneer*, November 14, 2013.
9. See http://radicalvalleys.wordpress.com/resistance-to-world-war-1-the-origins.
10. Stanbrook Abbey is a monastery of the Benedictine Order. Its nuns were interested in the research and the work of Ehrenfried Pfeiffer. They explored biodynamic practices and developed an organic garden (www.stanbrookabbey.org.uk).
11. Abtei Fulda *www.abtei-fulda.de.*
12. Kiyak, Mely, *Ein Garten liegt verschwiegen: Von Nonnen und Beeten, Natur und Klausur* [A Garden Lies Hidden: Of Nuns and Beds, Nature and Retreat], Hoffmann and Campe, Germany 2011.
13. Davenport, Andrew, *Quick Return Compost Making: The Essence of the Sustainable Organic Garden*, QR Composting Solutions, UK 2008, p. 29.

7. Ehrenfried Pfeiffer

1. Meyer, Thomas, *Ehrenfried Pfeiffer: A Modern Quest for the Spirit*, Mercury Press 2010, p. 60.
2. Ibid., p. 95
3. Ibid., p. 110.
4. Pfeiffer, Ehrenfried, *Unternatur und Übernatur in der Physiologie der Pflanze und des Menschen – die wahre Grundlage der Ernährung* [Sub-nature and Super-nature in the Physiology of Plant and Man: The True Basis of Nutrition], Dornach, 1958.
5. Saunders-Davies, Barbara, 'An Appreciation of Ehrenfried Pfeiffer', *Star & Furrow*, 1964.
6. Josephine Porter Institute for Applied Biodynamics, 201 East Main Street, Suite 14, Floyd VA 24091 www.jpibiodynamics.org.

8. Biodynamics in Post-War Britain

1. Mildred Kirkcaldy in the *Star and Furrow*, 1979.
2. George Corrin in the BDAA bulletin, July 1962.
3. Steiner, *Memory and Habit: Sense for Truth, Phenomenon of Metamorphosis in Life*, Anthroposophical Publishing Co., USA 1948, lecture of August 28, 1916. See also: Steiner, *The Riddle of Humanity*, p. 181f.
4. Koepf, Herbert, *Die Entwicklungsgeschichte der biologisch-dynamischen Landwirtschaft im 20 Jahrhundert* [The History of the Development of Biodynamic Agriculture in the Twentieth Century], Verlag am Goetheanum, Dornach 2001.
5. Mos, Uwe, *Die Wildgrasveredlung: Rudolf Steiners Impuls in der Pflanzenzucht* [Improving Wild Grasses. Rudolf Steiner's Impulse for Plant Breeding], Verlag am Goetheanum, Dornach 2006.
6. Jarman, Bernard, 'Alan Brockman', *Star and Furrow*, issue 119, summer 2013, p. 52.
7. Ibid.

9. Another Dimension

1. Müller-Wiedemann, Hans, *Karl König: A Central European Biography of the Twentieth Century*, Camphill Books, UK 1996.
2. König, Karl, *The Camphill Movement*, Camphill Books, UK 1993.
3. Rudolf Steiner's Soul Calendar verses translated by the author.
4. Count Nikolaus Ludwig von Zinzendorf (1700–60), founder of the Herrnhut community. His deeply held religious belief was that only through community could there be Christianity.
5. Amos Comenius (1592–1670) was a philosopher and educator and a bishop of the Moravian Brotherhood. His great ideal was the formation of a universal college where continuous learning could take place.
6. Robert Owen (1771–1858) was a social reformer and father of the cooperative movement.
7. Steiner, *Anthroposophy and the Social Question*, p.20.
8. See Karl König's *Social Farming*.

10. Community Supported Agriciulture

1. See www.stroudcommunityagriculture.org.

12. A Challenging Start to the New Millennium

1. Kolisko, Eugen and Lili, *Agriculture of Tomorrow*, p. 356.
2. Steiner, *From Comets to Cocaine* (CW348), p. 228.

13. Towards a Biodynamic Future

1. Section for Agriculture, *The Biodynamic Preparations in Context: Individual Approaches to Preparations Work*, Verlag am Goetheanum, Dornach 2016.
2. Paull, John, 'Biodynamic Agriculture, the Journey from Koberwitz to the World 1924-1938', *Journal of Organic Systems*, 6 (1), 2011.

Appendix 1: Office Holders of the Anthroposophical Agricultural Foundation (AAF)

1. Office in Gloucester Place, London.
2. A new constitution is approved and first members are enrolled.
3. Office moves to Old Mill House, Bray.
4. Margaret Cross and Leslie Binnie resigned.
5. Biological Research Institute moves to Bray with Lili Kolisko.
6. George Kaufmann changed his name to George Adams.
7. Carl and Gertrude Mirbt become Carl and Gertrude Mier.
8. Marna Pease retires as Secretary.

Appendix 2: Office Holders of the Biodynamic Association (BDA)

1. Nine-day Betteshangar conference in June 1939.
2. Biodynamic Association Trust founded in 1940 to hold property.

Appendix 3: Office Holders of the Biodynamic Agricultural Association (BDAA)

1. The role of Executive Director was initially, though not always, a paid role, whereas the role of Chair and the position of council members were voluntary charity roles.
2. A project to promote the development of biodynamic seeds was launched. Peter Brinch was employed as the Seed Development Officer. He sought to encourage seed production on farms and to develop the seed business in Botton.
3. Councils of BDA and AAF resign for new election.
4. Cynthia Chance starts newsletter.
5. From 1965 Council members no longer need to be members of the Experimental Circle.
6. From 1982 retiring Council members are ineligible to stand for re-election until a year has passed.
7. Rule change to allow Chair term of five years.
8. The Biodynamic Agricultural College (BDAG) was founded in 2008 to provide a new focus on BDA training programmes.

Bibliography

Bruce, Maye E., *Common-sense Compost Making By the Quick Return Method*, Faber & Faber, UK 1973.

Groh, Trauger and McFadden, Steve, *Farms of Tomorrow: Community Supported Farms, Farms Supporting Communities,* Biodynamic Farming and Gardening Association, USA 1990.

—, *Farms of Tomorrow Revisited: Community Supported Farms, Farms Supporting Communities*, Biodynamic Farming and Gardening, USA 1997.

Keyserlingk, Adalbert Graf von (ed.), *The Birth of a New Agriculture: Koberwitz 1924 and the Introduction of Biodynamics*, (J. Wood, trans), Temple Lodge, UK 2009.

Kolisko, Eugen and Lili, *Agriculture of Tomorrow*, Kolisko Archive Publications, USA 1978.

Meyer, Thomas, *Ehrenfried Pfeiffer: A Modern Quest for the Spirit*, Mercury Press, UK 2010.

König, Karl, *Social Farming: Healing Humanity and the Earth*, Floris Books, UK 2014.

Paull, John, 'The Betteshanger Summer School: Missing Link between Biodynamic and Organic Farming', ResearchGate, 2014.

—, 'Lord Northbourne, the Man Who Invented Organic Farming: a Biography', *Journal of Organic Systems*, 9(1), 2014.

—, 'The Pioneers of Biodynamics in Great Britain: From Anthroposophic Farming to Organic Agriculture (1924-1940)'. *Journal of Environment Protection and Sustainable Development*, 5(4), 2019.

Steiner, Rudolf, *Agriculture Course: The Birth of the Biodynamic Method* (CW327), Rudolf Steiner Press, UK 2012.

—, *Anthroposophy and the Social Question*, Mercury Press, USA 2006.

—, *The Evolution of Consciousness as Revealed Through Initiation Knowledge* (CW227), Rudolf Steiner Press, UK 2006.

—, *From Comets to Cocaine: 18 Lectures to Workers 1922-23*, Rudolf Steiner Press, UK 2000.

—, *Initiation Science and the Development of the Human Mind* (CW228), Rudolf Steiner Press, UK 2016.

—, *The Riddle of Humanity: The Spiritual Background of Human History* (CW170), Rudolf Steiner Press, UK 1990.

Woods, Abigail, *A Manufactured Plague: The History of Foot and Mouth Disease in Britain*, Earthscan, UK 2004.

Picture Credits

Public domain: Figures 2.2, 6.2.

Budd, David: Figure 7.2
Deimann, Gotz (ed), *Die anthroposophischen Zeitschriften von 1903 bis 1985*: Figures 3.4, 7.1.
The Dottendfelder Agricultural School: Figure 11.1.
Fyfe, Agnes, *Moon and Plant:* Figures 5.6a and 5.6b.
Jarman, Bernard: Figures 1.4, 3.6, 3.7, 5.2a–d, 5.3, 5.5, 8.1, 8.2, 8.3, 8.5, 9.2, 10.1, 10.2. 11.2.
Kaye, Gabriel: Figures 8.6 and 8.7.
Marcel, Christian, *Sensitive Crystallization:* Figures 7.3a and 7.3b.
Mier, Dorothea: Figure 3.3.
Paull, John: Figures 3.5 and 6.1.
Pietzner, Cornelius (ed), *The Candle on the Hill*: Figure 9.1.
Schöffler, Heinz Herbert, *Das Wirken Rudolf Steiners 1917–1925*: Figures 1.1, 1.2, 1.3, 2.1, 3.1, 3.2, 3.8, 5.1, 5.4.
Swann, Richard: Figures 8.4 and 8.8.

Index

Floris
Books